Information Systems for Engineering and Infrastructure Asset Management

Abrar Haider

Information Systems for Engineering and Infrastructure Asset Management

Foreword by Professor Andy Koronios

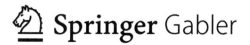 Springer Gabler

RESEARCH

Abrar Haider
Adelaide, Australia

Dissertation University of South Australia, Adelaide, Australia, 2007

ISBN 978-3-8349-4233-3   ISBN 978-3-8349-4234-0 (eBook)
DOI 10.1007/978-3-8349-4234-0

The Deutsche Nationalbibliothek lists this publication in the Deutsche Nationalbibliografie;
detailed bibliographic data are available in the Internet at http://dnb.d-nb.de.

Springer Gabler
© Gabler Verlag | Springer Fachmedien Wiesbaden 2013

Cover design: KünkelLopka GmbH, Heidelberg

Printed on acid-free paper

Springer Gabler is a brand of Springer DE. Springer DE is part of Springer Science+Business Media.
www.springer-gabler.de

To my mother

Our season of memories was way too short, just a life time!

Foreword

Contemporary engineering environments demand continuous availability and reliability of critical asset equipment, and engineering infrastructure utilised in manufacturing and community service. It is, therefore, not surprising to see the technological innovations and improvements relating to asset operation. These technological advances have improved the design, operation, and efficiency of these assets, and at the same time are transforming the way their lifecycle is managed. Conventional practices to manage assets are fast being revamped with the novel use of information technologies, and these technologies are expected to enable information driven integrated solutions aimed at preserving integrity of design and value profile of assets to their original or as designed specifications throughout their lifecycle. Industry is currently investing heavily in research activity finding solutions to achieve these objectives, though the existing lack of convergence of engineering and business management information relating to asset lifecycle is proving to be a major stumbling block. Engineering enterprises, therefore, are forced to follow a deterministic strategy to information systems implementation, where managerial emphasis is centred on implementation of technology rather than the causes and effects that shape the use of technology. Consequently asset managers have struggled with the implementation, use, and institutionalisation of information systems in order to translate the vast amount of available data into meaningful management information to optimise control and management of their asset base.

This research concludes that business value from information technologies adoption depends upon the organisational intent, interpreted through the strategic choices; and organisational context, shaped by the mutual interaction of various sub organisational institutions. However, owing to a deterministic view of technology, engineering enterprises treat information systems as passive technology constructs, whose behaviour is predicable and are able to provide same level of service regardless of their context of implementation. This text strongly posits that information systems are not objective entities, which could be implemented without considering their interaction with cultural, human, organisational, and technical factors. In fact, use and institutionalisation of these systems evolves over the course of time through continuous interaction of social, technical, and organisational contexts. Information systems implementation should, therefore, aim to improve organisational responsiveness to external and internal challenges by aligning these systems with information needs of the business.

This book draws on various theoretical perspectives relating to engineering, information technology, business governance, engineering management, and strategic business management paradigms, and focuses on implementation, use, and institutionalisation of information systems for engineering and infrastructure asset management. It serves as an excellent resource for asset managers to align information systems with strategic asset management information needs; provide integration of asset lifecycle processes to enable an integrated asset lifecycle management paradigm; and performance evaluation, control, and management of information systems based asset management. I am sure this book will introduce the readers to new perspectives on information systems, and the tools presented in this book will assist asset managers in better planning, execution, control, and management of asset lifecycle.

Professor Andy Koronios

Preface

Engineering and infrastructure assets maintain the lifeline of economies. It is, therefore, critical to manage these assets in such a way that they provide same level of service throughout their lifecycle. However, asset lifecycle management is information intensive and utilises a plethora of information systems. The role of these systems in asset management is much more profound and extends well beyond the organizational boundaries and addresses business relationships with external stakeholders to deliver enhanced level of business outcomes. In doing so information systems are required to translate strategic business considerations into action, and are also expected to produce learning and feedback that inform business strategy and aid in strategic reorientation. This, however, can only be attained if information systems infrastructure is aligned with the information requirements of asset lifecycle and facilitate informed decision support that acts as strategic advisory mechanisms supporting business planning and management.

This book is motivated by the lack of appropriate theoretical support for the implementation, use, and institutionalisation of information systems for asset management and evaluation mechanisms to measure the contributions that these systems make towards effectively managing the asset lifecycle. This book argues that information systems utilised for asset management are social systems, therefore, their use is socially composed and culturally institutionalised. Implementation and institutionalisation of implement these systems requires understanding the context of their implementation; the processes that influence the utilisation and implementation of these systems; and the organisational, cultural, people, and behavioural aspects influenced by their use. This book, therefore, adopts a multi-perspective context based approach and addresses operational as well as conceptual limitations of information systems implementation for asset lifecycle management. It entails taking stock of existing hard as well as soft organisational factors that enables actionable learning that acts as an advisory mechanism aimed at continuous improvement of information systems infrastructure for planning, execution, and management of asset lifecycle. This approach helps the organisation mature technologically by improving its responsiveness to the internal and external challenges.

This book provides in-depth analysis of the information systems utilised for asset lifecycle management by three Australian engineering infrastructure asset managing organisations. Based on the learnings from these studies, this book provides a new cognitive perspective to the knowledge of information systems based asset management. In doing so, it makes three major contributions to literature and practice of asset management. Firstly, it provides theoretical support for context based information driven information systems implementation and management, which accounts for the role of information systems as strategic enablers as well as strategic translators within the asset managing organisation. It seeks alignment of organisational sub-institutions with strategic asset management considerations through progressive maturity of information systems, so that they evolve with the organisation and mature in response to internal and external challenges posed to the organisation. Secondly, it presents a generative learning oriented information systems based asset management evaluation framework that facilitates continuous improvement of the asset management enabling information systems infrastructure. Thirdly, this research offers a comprehensive set of organisational, technical, and cultural factors that influence institutionalisation of information systems for engineering asset management. The contributions of this research to the field of asset management, equip asset managers with holistic understanding and knowledge of information systems utilisation to effectively plan, control, and manage asset lifecycle.

Dr. Abrar Haider

Table of Contents

List of Figures

List of Tables

List of Acronyms and Abbreviations

ABC	-	Activity Based Costing
ARMS	-	Asset Register and Management System
BOOT	-	Build Operate Own and Transfer
CBA	-	Cost Benefit Analysis
CE	-	Control Engineering
CMMS	-	Computerized Maintenance Management Systems
COAG	-	Council of Australian Governments
EAM	-	Engineering Asset Management
ERP	-	Enterprise Resource Planning
FMECA	-	Failure Mode, Effects, and Criticality Analysis
FTA	-	Fault-Tree Analysis
HAZOP	-	Hazard Operability Analysis
IRR	-	Internal Rate of Return
IS	-	Information Systems
IT	-	Information Technology
JIT	-	Just in Time
LCCA	-	Life Cycle Costing Analysis
LORA	-	Level of Repair Analysis
MRO	-	Maintenance, Repair, and Operations
MTA	-	Maintenance Task Analysis
NPV	-	Net Present Value
NUDIST	-	Non-Numerical, Unstructured Data Indexing, Searching and Theorizing
OH&S	-	Occupational Health and Safety
RA	-	Reliability Analysis
RCM	-	Reliability Centred Maintenance
RE	-	Reliability Engineering
ROI	-	Return on Investment
RTA	-	Roads and Traffic Authority
SCADA	-	Supervisory Control and Data Acquisition
TPM	-	Total Productive Maintenance

1 Introduction

Information Technologies are fast becoming the prime enabler of survival and success in business organisations. These technologies, on one hand, enrich economic, social, and cultural environment of organisations, and on the other hand enhance their competitiveness. A critical aspect of Information Technology (IT) adoption in general and Information Systems[1] (IS) in particular, is to find the strategic fit between the way an organisation executes its business and the technologies selected to aid in its execution. Organisations need to proactively seek to align their strategic considerations with information systems so that they become responsive to internal and external challenges. These systems should not be viewed as technical constructs or information deposits; in fact they are social systems involving people and are embedded in human organizations (Luna-Reyes, Zhang, Gil-Garcia, and Cresswell 2005; Lapiedra, Alegre, and Chiva 2006). Success of these technologies, however, depends upon the maturity of the technical, human resources, and organisational dimensions of the organisation. The variety of systems and the range of objectives associated with these systems, therefore, demand that organisations need to take stock of their capabilities, resources, and aspirations to enable informed choices regarding information systems implementation. However, a study by Australian Government's Department of Communications Information Technology and the Arts concluded that less than a third of all respondents had any post or pre implementation performance evaluation mechanism for investments in IT. Well over half the respondents reported that they never had such an item on their strategic agenda (DCITA 2005).

An attempt to implement information systems should be aimed at understanding the context of their deployment, as well as the processes that affect and are affected by their use (Stockdale and Standing 2006). Evaluation of these technologies, therefore, means assessments of hard quantifiable benefits that appear on an organisation's financial statements, as well as soft qualitative benefits that are reflected in organisational culture, behaviour, and intellectual capital (Frisk 2007).

This research aims to present an appreciation of the alignment of information system with strategic Engineering Asset Management (EAM) considerations, and the evaluation of information systems based asset management. In-

---

[1] Information Systems in this book refer to hardware, software, communication networks, and information systems that acquire, process, store, and deliver information to external and internal stakeholders in order to facilitate business processes. However, this thesis uses the terms IT and IS interchangeably.

formation systems are an integral part of asset lifecycle management. They perform various tasks at each stage of the lifecycle through data acquisition, processing and manipulation operations. However, the scope of information systems in engineering asset management extends well beyond these usual data processing, communication, and reporting capabilities and reaches out to business value chain integration, enhancing competitiveness, and transformation of business relationships (Haider, Koronios and Quirchmayr 2006).

1.1   IT and Engineering Asset Managing Organisations

Investments in IT have been increasing steadily in all industry sectors. According to Australian Bureau of Statistics (ABS 2007), during the year ended June 2006, engineering industry ranked highest among the IT enabled Australian businesses. More than 95% businesses of electricity, gas and water supply industries were using IT to enable their business processes, which was way higher than the collective Australian industries' average of 85%. However, construction industry lagged behind with a percentage of nearly 80%. Overall, Investments in IT have also been increasing steadily, with the government and private sector investing on average $26 – $27 billion since 2002 (ABS 2005a). This figure accounts for about 15% of annual total IT investments in Australia from the year 2002 till year 2005 (out of these figures government administration and defence contributed 12% and manufacturing industry contributed 9%) (ABS 2006).

In a survey conducted by Gomolski et al. (2001), it was concluded that IT investments in an organisation ranges between 1% and 3% of the total revenue. This figure, however, reaches 5% in service industries and thus outclasses expenditure on research and development activities. A study conducted by OECD (2006) reported that the IT investment divide between US and the rest of the world has been diminishing fast. IT investment in Europe and Japan were expected to grow at the rate of 2.2% and 2.8% respectively in 2006, whereas smaller OECD economies (Australia, Canada, Ireland, Korea, Mexico) and eastern European OECD countries (Czech Republic, Hungry, Poland, Slovak Republic) were all projected to have growth rates above the average OECD in 2006-07. This trend of augmented investments in IT shows that IT is increasingly being regarded as capital investment by businesses rather than operating expenditure (see for example Bajaj and Bradley 2005; Serafeimidis and Smithson 2003; Farbey et al. 1993).

Research in information system evaluation has gained momentum since the early 1990s. Although it appears that the value gained from IT investments vary substantially between industry sectors (OECD 2006; MGI 2002). Evaluation of information systems helps organisations in achieving a range of benefits, for example, mapping technology with business process needs, making informed choices about new investments in technology, enhancing organizational learning, and improvements in business integration and performance (Smithson and Hirscheim 1998; Farbey *et al.* 1999; Irani and Love 2001).

Engineering organisations traditionally take a deterministic view of technology while adopting information systems (Haider and Koronios 2005). It is for the same reason that investments in information systems carry the expectation of high future returns in terms of process efficiency and manufacturing/production/service provision output. Asset managing organisations, however, have twofold interest in information systems, first that they should provide a broad base of consistent and logically organised information concerning asset management processes; and second, the availability of real time updated asset related information available to asset lifecycle stakeholders (Rondeau *et al.* 2006). In turn, asset managers are looking for pragmatic asset and their lifecycle support solutions that exhibit solid proof of their value to the organisation. Information system departments, therefore, are becoming an integral part of strategic planning exercises for asset lifecycle management.

1.2 Asset Management

1.2.1 Defining an Asset

The term asset in engineering organisations is defined as the physical component of a manufacturing, production or service facility, which has value, enables services to be provided, and has an economic life greater than twelve months (IIMM 2006). Some examples include, manufacturing plants, roads, bridges, railway carriages, aircrafts, water pumps, and oil and gas rigs. Oxford Advanced Learner's Dictionary describes an asset as valuable or useful quality, skill or person; or something of value that could be used or sold to pay off debts (OALD 2005). These two definitions imply that an asset could be described as an entity that has value, creates and maintains that value through its use, and has the ability to add value through its future use. This means that the value it provides is both tangible and intangible in nature. A physical asset should be taken as an economic entity that provides quantifiable economic benefits, and has a value profile (both tangible and intangible) depending upon the value statement that its stakeholders attach to it during each stage of its lifecycle (Amadi-Echendu 2004). Management of assets, therefore, entails preserving the value function of the asset during its lifecycle including the economic benefits. Consequently, asset management processes are aimed at gaining and sustaining value from design, procurement and installation through operation, maintenance and retirement of an asset, so as to keep them as close to their as designed condition as possible (Blanchard and Fabrycky 1998).

Asset management represents a mix of strategic and integrated set of processes to gain greatest lifetime effectiveness, utilisation and return from assets (Mitchell and Carison 2001). According to Hastings (2000), asset management is derived from business objectives and represents set of activities associated with asset need identification, acquisition, support and maintenance, and disposal or renewal, in order to meet the desired objectives from asset operation effectively and efficiently. Fundamental aim of asset management is the continuous availability of value that it enables to its stakeholders through its service, production, or manufacturing provision. Consequently, asset management processes interact with a variety of other business processes within the organisation as well as with business partners, to allow for activities such as demand management, procurement, logistics, maintenance and repairs, and customer relationship management. Asset management, thus, represents a set of disciplines, methods, procedures and tools derived from business objectives aimed at optimising the whole life business impact of costs, performance and risk exposures associated with the availability, efficiency, quality, longevity and regulatory/safety/environmental compliance of an organisation's assets (Woodhouse 2001). This definition suggests that the scope of asset management processes is aimed at three levels, i.e. operational, management, and strategic. Operational level represents the set of activities necessary to keep the asset up and running to meet the stakeholders' needs; management level plans and manages how decisions taken at the strategic level are interpreted and implemented at the operational level; and the strategic level represents a long term focus on asset management from a total cost of ownership perspective (Sardar *et al.* 2006). In crux, asset management is policy driven, information intensive, value adding, and is aimed at achieving cost effective peak asset performance. The core objective of asset management processes is to preserve the operating condition of an asset to near original condition.

1.2.2   Principles of Asset Management

According to Stapelberg (2006), principles of asset management common to most organisations are that,

a. asset management is driven by corporate/business objectives and goals
b. assets exist to meet some identifiable service delivery, and asset management provides clearly assigned managerial accountabilities and responsibilities for service delivery,
c. asset management provides effective recognition and life cycle management of risks,
d. asset management places emphasis on optimising existing asset utilisation and performance to maintain or reduce overall service delivery cost to a sustainable level,
e. asset management considers the use of engineering solutions and management techniques for organising, planning and controlling the acqui-

sition, use, care, refurbishment, and/or disposal of an organisation's physical assets,

f. asset management takes a total life-cycle cost approach and utilises whole of life cycle techniques for assets to optimise their service delivery potential and to minimise the related risks and costs over their entire life, and

g. asset management focuses on performance measurement, monitoring and matching assets to organisation's strategic priorities.

1.3 Issues with IS based Engineering Asset Management

1.3.1 Evaluating Performance of Information Systems

Information systems implementation is often difficult (Smithson and Hirscheim 1998) and a wicked problem (Farbey *et al.* 1999), mainly due to the variety of roles that these systems have in the organisation. Evaluation by nature is a subjective term and is defined in the Oxford Advanced Learner's Dictionary as, the process of judging or forming an idea of the amount, value, or worth of an entity (OALD 2005). Evaluation of information systems is the assessment of their worth to the strategic objectives of an organisation. It is a process that is tightly coupled with other management and decision making processes. The connection between performance evaluation and strategic business adjustment is important; however, the knowledge required to evaluate performance is critical (Nonaka 1991). This knowledge allows the organisation to investigate and highlight the gap between the actual and desired performance. In order for that to happen it is essential that the organisation has access to relevant performance data. However, 70% to 90% of the organisations fail to realize success from their evaluations due to lack of availability of relevant data (Kaplan and Norton 2004).

1.3.2 Dynamics of Information Systems for Asset Management

Having its origin in mass production aimed at capturing market share, quality management calls for standardization of business processes managed by data and facts that focus on certain targets set by informed choices. However, quality of these informed choices cannot be guaranteed in business areas where such positivist assumptions are not valid (see for example Kirkpatrick and Lucio 1995; Adams 1998). Engineering asset management is one such area, where business processes are carried out in unpredictable environments, with conflicting objectives, and function on basically non-market transactions. Therefore creation, acquisition, dissemination, reuse, and management of information has serious operational and financial implications for the organisation. The fundamental issue in management of engineering assets is not just the quality of converting input to output, but also the control of information and knowledge guiding it and the use and reuse of such information for decision support and enterprise wide planning and business execution. In simple words, the fundamental issue here is not only doing things right, but also to have information that guides about what are the right things to do.

A typical asset lifecycle starts at the time of designing the manufacturing or production system, and typically illustrates stages such as, asset commissioning, operation, maintenance, decommissioning and replacement (IIMM 2006). Market demand and supply dynamics drive assets' and services design, and this design drives production/manufacturing/service provision. This, in turn, specifies the operational workload of an asset. Operational workload and asset design generate maintenance demands to keep the assets in running condition, whereas maintenance determines the future production capacity of the assets as well as their remnant life span (Haider and Koronios 2003). However, information systems utilised in asset management not only have to provide for the decentralized control of asset management tasks but also have to act as instruments for decision support. For example, a critical aspect of effective asset lifecycle management is the learning or knowledge gained at each stage, which provides for the feedback to other processes. Asset operation profiling has significance for asset redesign as well as asset maintenance, asset operation cost benefit analysis, and lifecycle decision support (Haider and Koronios 2006). Information systems for engineering asset management, therefore, are required to enable an integrated view of lifecycle information such that whole lifecycle impact could be considered before making choices about asset lifecycle. This integrated view, however, requires appropriate hardware and software applications; quality, standardised, and interoperable information; appropriate skill set of employees to process information; the strategic fit between the asset management processes and the information system; and a conducive organisational environment.

Current information systems in operation within engineering enterprises have outlived their productive lives, as the methodologies employed to design these systems define, acquire and build systems of the past not for the future (Haider and Koronios 2004a). For example, the maintenance information system development that has attracted considerable attention in research and practices are far from being optimal. While maintenance activities have been carried out ever since advent of manufacturing; modelling of an all inclusive and efficient maintenance system has yet to come to fruition (Duffuaa *et al.* 2001; Yamashina 2000). This is mainly due to the continuously changing and increasing complexity of asset equipment, and the stochastic nature or the unpredictability of the environment in which assets operate, along with the difficulty to quantify the output of the maintenance process itself (Duffuaa *et al.* 1999). Current information systems employed for condition monitoring identify a failure condition when the asset is near breakdown, and therefore serve as tools of failure report-

ing better than instruments for pre-warning the failure condition in its development (Haider and Koronios 2004b).

In response to the increased competitive pressures, maintenance strategies
that once were run-to-failure are now fast changing to being condition based,
thereby necessitating integration of asset lifecycle information. This requires
integration of decision systems and computerized maintenance management systems in order to provide support for tasks such as maintenance scheduling, maintenance workflow management, inventory management, and purchasing (Bever
2000). However, in practice, data is captured both electronically and manually,
in a variety of formats, shared among an assortment of off the shelf and custom-
ized operational and administrative systems, communicated through a range of
sources and to an array of business partners and sub contractors; and consequently inconsistencies in completeness, timeliness, and inaccuracy of information leads to the inability of quality decision support for asset lifecycle management (Haider and Koronios 2005). In these circumstances, existing asset management information systems could be best described as pools of isolated data
that are not being put to effective use or to create value for the stakeholders.

1.3.3   IT Investments and Productivity Paradox

Owing to a deterministic view of technology, managerial expectations from IT
investments are those of increased quality and quantity of output, as well as substitution of human effort through automation (Parker *et al.* 1997). These expectations also contribute to the underlying assumptions relating to IT investments
that their adoption will outdo related costs. Advantages of these costs benefits
are often translated as gains in terms of production/manufacturing/service provision output through operational efficiency. However, research suggests that the
assumptions about productivity gains from IT investments are paradoxical. Despite substantial investments organisations, industries, and even national economies have failed to register an increase in their productivity. This phenomenon
has been termed as productivity paradox (Brynjolfsson 1993). Value from IT,
however, is contingent upon the maturity of the organisational and human aspects with technology.

Generally, engineering enterprises mature technologically along the continuum of standalone technologies to integrated systems, and in so doing aim to
achieve the maturity of processes enabled by these technologies, and the skills
associated with their operation (Haider *et al.* 2006). Konradt *et al.* (1998) further
assert that engineering enterprises adopt a traditional technology-centred approach to asset management, where technical aspects command most resources
and are considered first in the planning and design stage. Skills, process maturity, and other organisational factors are only considered relatively late in the
process, and sometimes only after the systems are operational. However, human,
organisational, and social factors have a direct relationship with information sys-

tems (Checkland 1981; Orlikowski and Robey 1991; Orlikowski and Baroudi 1991; Walsham 1993, 1995), which underscore the conceptual and operational constraints posed to effective information system implementation.

1.3.4   Conceptual Limitations of Evaluating IS based EAM Value Profile

Evaluation, conceptually, is a subjective activity that is biased and cannot be detached from human understanding, social context, and cultural environment, within which it takes place. Evaluation, therefore, is influenced by the actors who carry out this exercise; and the principles and assumptions that they employ to execute evaluation. Considering the fact that human interpretation shapes and reshapes with time, the nature of evaluation also changes from time to time. Evaluation, thus, represents the existing meanings and interests that individuals or communities associate with the use of technology within the socio technical environment of an organisation. The focal point of socio technical perspective is the interactive association between people, information system and the social context of the organisation (Bijker *et al.* 1987; Orlikowski 1992; Bijker and Law 1992). However, action is an important element of this interaction. This notion of action is contained in the structuration theory (Giddens 1984), which posits that action is facilitated and influenced by the social structure. People's interaction with technology is fashioned by the social structure and their actions persistently shape or transform social structure (Hayes and Walsham 2000). There is, therefore, a dynamic relationship between technology, the context within which it is employed, and the organisational actors who interact with technology. This duality of technology is characterised by Orlikowski (1992), who argues that technology is socially and physically constructed by human action.

When technology is physically adopted and socially composed, there is generally a consensus or accepted reality about what the technology is supposed to accomplish and how it is to be utilized (Bijker *et al.* 1987). This temporary interpretation of technology is institutionalised and becomes associated with the actors that constructed technology and gave it its current significance (Orlikowski 1992), until it is questioned again for reinterpretation. This requirement of reinterpretation may grow owing to changes in the context, or the learning that may render the current interpretation obsolete. Technology, therefore, is not an objective entity, such that it could either be evaluated without considering its interaction with social and human factors (Manion and Evan 2002), or it could be evaluated in basic and one-dimensional economic terms (Bjorn-Anderssen 1988; Orlikowski 1992; Sauer and Yetton 1997; Truex, Baskerville, and Klein 1999; Atkins and Dawson 2001).

When information system evaluation is employed it is expected that it will expose a number of different dimensions of information system implementation, such as, financial, technical, behavioural, social, and management aspects of information system. Furthermore, these endeavours may be aimed at stakeholder satisfaction, role of information system, and information system lifecycle. These expectations change during the lifecycle of an information system. An ex ante or pre implementation evaluation is aimed at ascertaining cause and effect of technology; whereas, ex post or post implementation evaluation may be aimed at evaluating how well the organisational information systems translate strategic business objectives into action by enabling, automating, and integrating business processes; as well as how well these systems inform business strategy with the quality of their information analysis and decision support. Each of these dimensions, their related objectives and aims have their own theories, postulates, and evaluation criteria, which makes a comprehensive information system evaluation complicated and extremely difficult.

1.3.5   Operational Limitations of Evaluating IS based EAM Value Profile

Contemporary asset management paradigm demands an elevated ability and knowledge to incessantly enable asset management processes, with support in terms of quality data acquisition, real-time data exchange, and computer supported categorization and analysis of asset's operational divergences from standard procedures (Sandberg 1994). Bamber *et al.* (1999) argue these factors are essential for effective planning, scheduling, monitoring, quality assurance, and acquisition of necessary resources required for supporting asset lifecycle, and consequently enhancing the competitive profile of the asset managing organisation. Role of IT investments is no more considered as inwardly looking systems aimed at operational efficiency through process automation; in fact, it extends beyond the organisational boundaries and also addresses areas such as business relationships with external stakeholders to deliver desired business outcomes. This complicates the process of decision making for IT investments, since this decision needs to address the impact of technology on business processes and resources, as well as integration of these technologies with existing technical and organisational infrastructure. However, information system evaluation generally has a narrow focus and involves people who cannot evaluate IT on anything other than technological dimensions (Willcocks and Lester 1997). Consequently, simplistic measures are adopted to measure the effectiveness of information systems, where the efficacy criteria are aimed at process efficiency rather than the organisational transformation prospectus of these systems. The measurement attributes involved in such IT evaluations, require both aspects of IT benefits to be taken care of i.e. soft benefits, such as stakeholder satisfaction, and customer relationship management; and hard benefits, such as cost, and information system throughput. However, evaluation methods lacking in completeness render the accuracy and credibility of evaluation mechanisms questionable.

Information system evaluation has different objectives and aims ex ante and ex post. In ex ante or pre implementation technology evaluation, decisions are generally based on cost benefits, and the perceived value that the investment may bring to the organisation. This investment is usually carried out by functional teams, who evaluate different choices of technologies and then arrive at a decision. The measurement criteria are often not clear and basically governed by the assumptions of the future use of technology, as conceived by the evaluators. On the contrary, during a post implementation evaluation a report card on the performance of information system is developed. This type of evaluation is generally not conducted by people who conduct ex ante evaluations, and therefore susceptibilities of technology in terms of purpose, and effectiveness of use are not considered. Furthermore, these two factors change with time, mainly due to technological innovation and changes in business environment. Post implementation evaluation is also expected to produce learning and feedback that could be used for strategic reorientation. However, this form of evaluation requires long term involvement, and experience, such that the purpose, use, and fit of technology within the organisation are understood. This makes the success or failure of information system open to interpretation according to the judgements and experiences of the evaluators.

Information system evaluation calls for ascertaining both hard as well as soft benefits to the organisation by using quantitative as well as qualitative means and their connection to organizational development (Farbey *et al.* 1993; Grembergen and Bruggen 2003). This can only be attained if information system evaluation provides a roadmap in terms of alternatives and choices (Apostolopoulos et. al 1997; Fasheng and Teck 2000), and hence becomes a strategic advisory mechanism that supports planning, decision making, and management processes (Hawgood and Land 1988; Karlsson and Gennas 2005). Such evaluations provide feedback (Simmons 1996; Serafeimidis and Smithson 2003) that facilitates organizational learning (Nevis *et al.* 1995; Argyris and Schon, 1996; Farbey *et al.* 1999) and indicates the fundamental reasons, factors, and causes for underperformance or success of IT investments (Davern and Kauffman 2000).

1.4   Research Background and Research Questions

1.4.1   Need for Research

Organisations' expectations associated with adoption of IT are quite diverse, such as operational efficiency, reduction in operating expenses, or enhanced competitiveness. However, there are divergent views held about the value creation of IT investments. Although, recent studies have concluded that IT investments provide positive economic returns (Anderson *et al.* 2002; Brynjolfsson and Yang 1999); nevertheless the impact of IT investments varies within organisations (Leibs 2002). Evidence found in literature, both industry and academic, sustains the argument of success (see for example, Devaraj and Kohli 2002; MGI 2001; Remenyi 1991) and failure (see for example, Ehrhart 2002). The reason for this polarisation is the propensity to neglect the active interaction and shared shaping between technology and people (MacKenzie and Wajcman, 1999; Silverstone and Haddon 1996). It is also argued that when organisations attempt to evaluate IT, managerial emphasis is mostly on improving cost profile of IT adoption. Majority of information system evaluation exercises are carried out using capital investment appraisal techniques, such as cost benefit analysis, payback and return on investment (Serafeimidis 1997; Serafeimidis and Smithson 2000). These evaluations only give a slice of the total impact of IT investments and disregard the human and organisational aspects of IT adoption, and, therefore, not only keep the softer benefits hidden but the costs of managing these benefits also remain unknown (Khalifa *et al.* 2001). Furthermore, these unobserved benefits prevent the users to realise the full potential of these systems (Pennington and Wheeler 1998). Consequently, such evaluations fail to measure the total impact of IT and contribute to failure of IT investments to achieve desired objectives (Pouloudi and Whitley 1997).

Liyanage and Kumar (2003) argue that the changing competitive environment of asset managing organisations along with stricter regulatory requirements, are forcing asset managing organisations to have effective performance management mechanisms for their asset management processes. This trend is getting popular in capital intensive industries, such as petroleum (Dwight 1999; Tsang 1999; Liyanage and Kumar 2000). As these industries are increasingly becoming aware of the shortcomings of the classical measurement techniques that focus on financial aspects of evaluation (Kaplan and Norton 1996; Sveiby 1997), asset managing organisations like BP, Shell, and Phillips are broadening the scope of their evaluation exercises, so as to include soft as well as hard determinants of engineering asset management (Liyanage and Kumar 2003). However, even though evaluation mechanisms based on financial measures alone have not yielded appropriate evaluations for organisations, yet the trend of using these techniques continues (Ballantine and Stray 1998). That is why financial criteria for performance evaluation dominate measurement exercises. This may be attributed to the natural prudence of business managers while making investments decisions. Nevertheless, more research is required to determine the barriers and impediments that discourage more refined approaches to information system evaluation (Yusof *et al.* 2006; Andresen 2001).

This book argues that evaluation of information system based engineering asset management is an important aspect of IT investments management in engineering organisations. However, there are certain conceptual and operational issues and challenges that impede employment of an effective evaluation mechanism, which justify the need for research in this area. This book posits that information system based engineering asset management evaluation is not yet developed at theoretical level, and the techniques and methods employed in practice lack requisite features of an all encompassing evaluation. This research, therefore, investigates conceptual and operational dimensions of information system based engineering asset management evaluation. As has been mentioned earlier, information system evaluation is not an inert or stagnant activity; in fact it is highly influenced by the organisational environment. In terms of conceptual dimension, this research investigates an engineering asset management context based information system evaluation, by moving away from the traditional assumptions of information system evaluation. It calls for pluralism, which demands qualitative as well as quantitative measures, involving cultural, social, economic, political, technical, and organisational aspects. Since change is innate element of organisational dynamics, context based evaluation of information system becomes essential.

On the operational side, this research is motivated by the challenges and gaps faced by asset managing organisations with regard to IT adoption. As discussed earlier, asset managing organisations utilise a variety of information systems to support the lifecycle of their assets; however, these systems do not perform to their true potential. At the same time competitive and operational pressure require these organizations to manage the health and operational capacity of their assets just like humans do to safeguard against disease and loss of productivity. However, humans could be considered as active machines since they can process information, whereas assets are passive and are unable to process information at their own. A relevant dimension of an information system based integrated view of engineering asset management (as discussed in section 3.1) is to provide the assets with the ability to process information as humans do. However, this requires the strategic fit between asset lifecycle needs and right choice of information system technology, which in turn requires effective evaluation methods that allows the information system based asset lifecycle processes to 'speak out their needs'.

1.4.2 Research Scope and Questions

The main aim of this research is to develop an appreciation of the evaluation methods for information system based engineering asset management. It works on the preposition that in response to information system implementation, certain changes have to take place in the organisational and social environment of an organisation. Nevertheless, this research does not deal with the organisational and social features that support institutionalisation of information systems. The objectives of this research are to understand, how information systems provide value to engineering asset management; and how to evaluate these systems. Core objective of this research are to understand how asset managing organisations utilise information systems for asset lifecycle management, and how could information systems based asset management efficiency be evaluated. It can be viewed as a feedback embedded arrangement that builds on the changes brought about by information system evaluation, the way information systems are institutionalised in the organisation, and recognizes information system adoption as a strategic enabler as well as strategic translator. Thus, the scope of this research deals with understanding the dynamics that shape institutionalisation of information systems for engineering asset management, and evaluation methods of information systems based engineering asset management. The research questions addressed in this research is,

Information systems allow for an integrated view of asset lifecycle management. However, why do asset managing organisations generally fail to evaluate the performance of information system based asset lifecycle processes, which could enable them to better understand and manage the performance and needs of asset lifecycle?

This question requires responding to the following three sub questions,

SQ 1:   *How do information systems facilitate alignment of strategic asset management considerations with overall business strategy and organisational design?*

SQ2:   *What factors impact institutionalisation of information systems based engineering asset management processes and their performance evaluation?*

SQ3:   *How information systems based asset lifecycle management processes should be evaluated?*

A few aspects of these questions require explanation. Firstly, in the main research question the term, 'fail', refers to both, failure to employ an evaluation method at all, and failure to achieve requisite results from an evaluation method. Secondly, the term 'facilitate' in sub question 1. Information systems are not passive entities, and their effective implementation requires appropriate organisational environment for them to perform at their true potential. In the first sub question, therefore, this research investigates how information systems allow alignment of strategic asset management considerations with overall business strategy and organisational design, and what factors influence and are influenced by their adoption?

1.4.3   Research Methodology

This research employs an interpretive epistemology with a qualitative perspective. It is obvious that the issues relating to evaluation of IT investments in engineering asset management are complex and multifaceted, and require a broad and flexible perspective for comprehensive examination. These include technical issues as well as an assortment of organisational, social, and cultural issues. Information systems can be classified as interpretive if it is assumed that our knowledge of reality is gained only through social constructions such as language, consciousness, shared meanings, documents, tools and other artefacts. Interpretive research does not predefine dependent and independent variables, but focuses on the complexity of human sense making as the situation emerges (Kaplan and Maxwell 1994). It attempts to realize the phenomena under investigation through the meanings that people attach to them (Deetz 1996; Boland 1991; Orlikowski and Baroudi 1991).

In order to address the research questions an interpretive stance provides a richer understanding of the context of information system based engineering asset management issues than the more conformist positivist approaches. Klein and Myers (1999) propose, seven principles for conducting interpretive research in information system, these are, the fundamental principle of the hermeneutic circle; the principle of contextualization; the principle of interaction between the researchers and the subjects; the principle of abstraction and generalization; the principle of dialogical reasoning; the principle of multiple interpretations; and the principle of suspicion. By applying these principles in general and the principle of contextualization in particular, this research examines information systems adoption in engineering asset management context; how are information systems interpreted and adopted by humans in their jobs; how engineering asset management supported by information system affects and is affected by organisational dynamics; and what theories could be produced through these explorations to better understand the nature of information systems based engineering asset management evaluation.

The research strategy adopted for this research is multiple exploratory case studies in Australian asset managing organisations. In interpretive research, case studies examine the details, connotations, and meanings of experience and do not generally prove or examine a hypothesis. In fact, the researcher undertakes to identify and uncover patterns and themes in data. Exploratory case study as a research strategy has increasingly been used in information system research (Klein and Myers, 1999; Orlikowski, 1996; Walsham, 1995). Case study allows the researcher to examine a phenomenon in its real-life context (Yin 1994) by making use of multiple methods of data gathering from a range of entities. Benbasat et al. (1987) further assert that a case study can be used where the research and theory are at their early, formative stages. Given that little research has been conducted on information system based engineering asset management, there is a need to examine this phenomena from a real world perspective.

The research strategy is designed to emphasize the role of information system in engineering asset managing organisations, and hence a narrative approach has been adopted to conduct interpretive field research within three organisations. The field work was conducted between August 2005 and February 2007. The methods employed in the field study include tools such as interviews, surveys, official documents, and direct observations. This research, therefore, employs triangulation principle that seeks to validate research results by coalescing variety of data sources (Tashakkori and Teddlie 1998). Denzin and Lincoln (1994a) argue that triangulation improves the probability of the acceptance of interpretation due to the supporting evidence available at each stage of data collection. Triangulation, thus, increases the probability of producing credible findings (Lincoln and Guba 1985).

1.5   Book Structure

There are eight chapters in the book. Chapter 1 provides the background to this research, launches the research problems, and presents the research questions for investigation. It also includes justifications of the research and a synopsis of the research approach and methodology.

Chapter 2 reviews literature on asset management, which leads to the understanding of asset lifecycle processes. From this review of the literature, a theoretical framework of asset management has been developed that provides the basis for discussion on information system based engineering asset management in chapter 3.

Chapter 3 critically reviews literature on information system implementation and the role of information systems in engineering asset management. This chapter investigates the barriers posed to information systems implementation in asset managing organisations, and discusses the essential elements of an integrated asset lifecycle view. It particularly discusses major information system implementation theories that provide theoretical foundation for information system implementation and thus helps in formulating a framework for alignment of information systems with strategic asset management.

Chapter 4 reviews literature on evaluation methodologies in general, and information system based performance evaluation in particular, and thus encapsulates the broad perspective of the research theme. It reviews the performance evaluation of the various dimensions and roles of information systems in contemporary business environment.

Chapter 5 describes and justifies the research methodology chosen for this research within the interpretive paradigm. It discusses the range of research approaches and methodologies available in information system research, and provides rationale of adopting interpretive approach to resolve the research questions of this study. It also describes the process of field study together with data collection, analysis, and reporting procedures.

Chapter 6 discusses the empirical data collected for this research. It provides in-depth description of three case studies conducted for this research. These case studies provide description of the case organisations; information system utilised for engineering asset management; problems and issues regarding information systems based asset management in case organisations; and the institutionalisation and maturity of information systems in each organisation. Case study results are illustrated through triangulated data sources, which include direct quotations of the interviewees to reinforce research findings.

Chapter 7 presents the analyses of the case study data. It answers the research questions by building on the within and cross case analyses that highlight factors that shape information systems based asset management and information systems based asset management evaluation, and issues that impact information systems for asset management.

Chapter 8 concludes this book. It highlights the contributions to this research to the body of knowledge and outlines the implications for theory and practice. The chapter also discusses limitations of this research and further research directions.

2 Asset Management

This chapter reviews literature to develop an understanding of the conceptual basis of asset management. In order to investigate information systems implementation for asset management and performance evaluation of information systems based asset management, it is essential to sketch out how asset lifecycle management processes are carried out, what are the demands of asset lifecycle management, what factors influence asset lifecycle management, and how asset managing organisations learn and adapt to changes in their business environment. This chapter sets the scene for later chapters by discussing in detail the fundamental characteristics of each asset lifecycle stage, their interrelationships, and their impact on overall asset management.

Since recent past, manufacturing and production environment is subjected to radical changes, which have been fuelled by intensely competitive liberalised markets, technological advances promising enhanced services, and improved asset infrastructure and plant performance. This emergent industrial re-organisation has a direct influence on economic incentives associated with the management of asset equipment and infrastructures, since continuous availability of these assets is crucial for profitability and efficiency of the business. As a consequence, engineering enterprises are faced with new challenges of safeguarding technical integrity of their assets, by coordinating support mechanisms required to keep them close to their original or as designed specifications and operational capabilities. However, there is no one size fit all approach to asset management, since it depends upon the nature of business and the types of asset that the business employs. Consequently, there are a variety of approaches available in research and practice, with each attempting to manage assets and resolve related issues in its own unique ways.

Chapter 1 suggested that asset lifecycle management represents a set of processes that are derived from the strategic goals of the organisation, and are designed to create value towards achieving these goals and objectives. This chapter establishes the link of asset lifecycle management processes to the strategic objectives of the organisation, and presents a comprehensive asset lifecycle management framework. Each aspect of this framework is discussed in detail by covering the core asset lifecycle management processes as well as asset reliability paradigms, asset lifecycle support processes, and asset management enabling infrastructure. After a thorough review of literature, it becomes clear that effective asset management requires an integrated approach that aligns asset lifecycle management processes with organisational (financial and non financial) and technological infrastructure.

2.1   Asset Management

The scope of asset management extends from establishment of an asset management policy and identification of service level targets as per requirements of stakeholders, to controlling daily operations of assets aimed at meeting the defined levels of service, to managing relationships with external businesses involved in maintenance and management of assets. Asset managing organisations, therefore, are required to cope with wide range of changes in the business environment. Asset management can be classified into three levels, i.e. strategic, tactical, and operational (Figure 2-1).

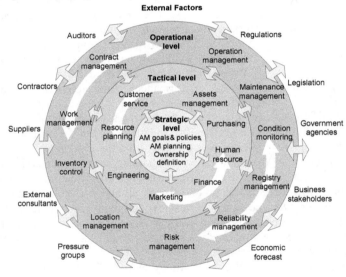

Figure 2-1: Scope of Asset Management
Source (IIMM 2006)

Strategic level is aimed at responsiveness to internal as well as external challenges. It is concerned with understanding the needs of stakeholders, market trends, and linking the requirements thus generated to the tactical/planning and operational level activities. Operational and tactical levels involve planning, decision support, monitoring, and review of each lifecycle stage to ensure availability, quality, and longevity of asset's service provision. The identification, assessment, and control of risk is a key focus at all levels, and the results thus achieved provide inputs to the asset management strategy, policies, objectives, lifecycle processes, operational and maintenance plans, controls, and management of resource.

2.1.1 Strategic Asset Management

Asset management has evolved from the humble beginnings of maintenance of plant and machinery and related functions (Sethi and Sethi 1990) to an approach that is equally as important and essential as total quality management aimed at reliability and efficiency of the organisation (Narain *et al.* 2000). Asset lifecycle strategic planning typically has a 10-25 year horizon for financial planning purposes, although organisations may look well beyond this period in order to fully assess the optimum lifecycle strategies (IIMM 2006). Strategic asset lifecycle planning translates legal and stakeholder requirements into service outcomes; thereby allowing for an overall long term vision of managing assets. The main features of strategic asset management planning include,

    a.   development of vision, mission and values statements which describe the long-term desired position of the organisation and the manner in which the organisation will conduct itself to achieve the same (Kerns 1999; Alexander 2003);

    b.   review of the operating environment, to ensure that all elements that have an impact on asset management are considered. Such elements include corporate, community, environmental, financial, legislative, institutional and regulatory factors (Angell and Klassen 1999; King and Lenox 2001; Rothenberg *et al.* 2001; Inman 2002);

    c.   identification and evaluation of strategic options to achieve strategic goals arising from the vision and mission statements (Narain *et al.* 2000; Boyle 2006); and

    d.   a clear statement of strategic direction, policies, desired outcomes, and risk management plans (Balch 1994).

Public sector organisations may give more weighting to environmental, social and economic factors in determining strategic goals; whereas private sector asset owners will typically place most emphasis on economic factors (IIMM 2006). However, it is the agreement on levels of service in terms of criteria such as quality, quantity, timeframes and cost that provides the link between strategic and tactical plans.

2.1.2 Tactical Asset Management

Tactical level planning involves the application of detailed asset management processes, procedures, and standards to develop sub-plans that allocate resources (both financial and non financial) to achieve strategic goals by meeting defined

levels of service. Depending on an organisation's purpose, tactical plans may
have varying priorities, for example, owners of infrastructure assets are usually
more (and in a much direct way) concerned with asset operations and management plans and customer service plans, which then provide the foundation for
other tactical plans, such as maintenance resources acquisition and management
plan. The fundamental aim of tactical asset management is to cost-effectively
achieve the organisation's strategic goals in the long-term. These plans, procedures, and standards cover asset management aspects, such as,

    a.   setting asset management objectives, including technical and customer
        service levels and regulatory/financial requirements (Foster *et al.* 2000;
        El Hayek *et al.* 2005);

    b.   operational controls, plans, and procedures (Lockamy 1998; Taskinen
        and Smeds 1999);

    c.   managing asset management information systems and information con-
        tained in them, such as asset attributes, condition, performance, capac-
        ity, lifecycle costs, maintenance history, etc. (Karababas and Cather
        1994; Gottschalk 2006);

    d.   risk management (Murthy *et al.* 2002; Balogun *et al.* 2004);

    e.   asset performance, health management, and condition assessments
        (Prickett 1999; Sherwin 2000; Tsang 2002); and

    f.   decision making for optimisation of asset lifecycle management
        (Lindberg 1990, Blanchard 1996).

2.1.3 Operational Asset Management

Operational plans generally consist of detailed implementation plans and information framework to enable these plans. These plans usually have a 1-3 year
outlook. These plans provide direction to the organisation on annual or biannual
basis and are concerned with practical rather than visionary elements. According
to IIMM (2006) operational plans typically include aspects, such as:

    a.   operational controls to ensure delivery of asset management policy,
        strategy, legal requirements, objectives and plans;

    b.   asset workload specification, condition monitoring, and process control;

    c.   structure, authority and responsibilities for asset management;

    d.   Staffing issues - training, awareness and competence;

    e.   consultation, communication, documentation; to/from external stake-
        holders, management, and employees;

    f.   information and data control; and

    g.   emergency preparedness and response.

Operational plans actually work as practical translations for priorities arising
from tactical plans in order to deliver cost effective levels of service. Various
researches stress inclusion of auditable performance measures (See for example

Gibb 2002; Tangen 2004) in these plans, so that effectiveness of these plans could be measured.

2.2   Asset Lifecycle

An asset lifecycle typically illustrates stages such as, asset commissioning, operation, maintenance, decommissioning and replacement. These stages represent an interesting mix of cause and effect and are interrelated, where market demand and supply dynamics drive product and services design, and this design drives asset need. This need definition specifies the operational workload of an asset, which in turn generates maintenance demands to keep the assets in running condition. Maintenance determines the future production capacity of the assets as well as their remnant life span. Table 2-1 below highlights the processes and the activities involved in an asset lifecycle.

Lifecycle Actions	Description	Focus
Strategic Planning	Planning of asset management functions, processes, technology, lifecycle costing, level of service, and support infrastructure to meet business goals, and asset demand management.	Strategic fit between overall business objectives and each activity performed in asset lifecycle management to enhance competitiveness of the businesses.
Core Asset Management Actions	Assembly/ procurement/ construction, operation, condition assessment, maintenance, refurbishment, replacement, and disposal of assets.	Ensuring availability, efficiency, quality, longevity & regulatory/ safety/ environmental compliance of asset operation at minimal costs with as designed specifications.
Asset creation/ acquisition	Capital investments to acquire, construct, or improve an asset to satisfy or improve level of service, stakeholders' demand management through manufacturing/ production/ service provision of an asset.	Asset reliability, compliance, availability through effective design, functional analysis, and economic tradeoffs; e.g. high development costs might be traded off against lower maintenance costs through improved reliability.
Asset Operation	Smooth asset operation through close collaboration with maintenance function, and providing feedback on asset operational behaviour to design and maintenance functions.	Efficiency and quality, regulatory/ environmental compliance of asset operation through conformance with planned operating conditions, as designed instructions, and environmental regulations.

Continued Next Page

Lifecycle Actions	Description	Focus
Asset Operation Assessment	Asset operation monitoring to assess the ability of asset to meet required service levels and mitigate operational risks, including continuous or periodic inspection, assessment, reporting, and interpretation of resulting information to indicate the condition of the asset in order to specify the nature and timing of maintenance.	To ensure asset longevity by proactively assessing asset condition in order to predict developing failure conditions so that maintenance could be planned.
Asset Maintenance	Sustain and where necessary perform necessary repair and maintenance on assets so that they continue to fulfil their functions and make the required value adding contribution to the manufacturing or production process, it includes actions such as maintenance planning, maintenance work execution, spare supply chain management, routine repairs, and testing.	Ensuring efficiency and longevity of asset by taking necessary actions to retain an asset to near original asset configuration/specification and condition by decreasing its rate of deterioration, thereby minimising asset downtime.
Asset Rehabilitation	Rehabilitation of assets through technology refresh, or maintenance; including actions such as treatments for improving asset condition and configuration to attain or exceed as designed state in order to technically update the asset to realign asset profile with changing strategic objectives.	Review of asset configuration and service delivery aspects, and their fit with the strategic business objectives.
Asset Disposal/ Retirement	Investments aimed at sustainable rationalisation through replacing or deconstructing an existing asset through end of need, in order to acquire or setup new assets to enhance level of service, and/or to improve configuration, and/or change in physical location.	Exploiting continuous improvement opportunities in better asset design and configuration to provide elevated level of service.
Asset Lifecycle Supportability Modelling	Modelling of asset design, construction, configuration, operation, maintenance, cost benefit analysis, and support infrastructure to ensure asset reliability, availability, and quality.	An integrated approach to performance tradeoffs and lifecycle decisions, through analytic models that predict asset behaviour; and changes in asset design, and operation; tradeoffs regarding individal stages of asset lifecycle.

Continued Next Page

Lifecycle Actions	Description	Focus
Technology Support	Combination of hardware; software; information acquisition, processing, communication and storage infrastructure; and skills to execute lifecycle processes.	Strategic fit between technology and asset management processes, aimed at availability of timely, consistent, complete, accurate, and reliable information.
Performance Assessment	Review and assessment of asset lifecycle management plans to measure their effectiveness in satisfying business needs.	Audit and assessment of implementation and execution of planned activities against documented standards, objectives, strategies, and stakeholder requirements aimed at continuous improvement of the asset management process.
Lifecycle Learning Management	Profiling asset operation, managing lifecycle knowledge, and feedback for better understanding of asset behaviour, and continuous improvements in asset design, operation, maintenance, reinvestment, compliance, and asset lifecycle support.	Combination, preservation, and use/reuse of lifecycle learnings and knowledge to provide an integrated view of the overall asset lifecycle management.

Table 2-1: Asset Management Lifecycle Management Framework (Developed from Husband 1976; Fabrycky and Blanchard 1991; Campbell 1995; Blanchard 1997; Woodward 1997; Kelly 1989; Blanchard and Fabrycky 1998; Moubray 2000; Woodhouse 2001; Mitchell and Carlson 2001; Waeyenbergh and Pintelon 2002; Eerens 2003; Moubray 2003; Amadi-Echendu 2004; IIMM 2006; Schuman and Brent 2006).

2.3   Strategic Planning

Porter (1979; 1980; 1996) proposes the concept of value chain and argues that business activities must be organised as such that they collectively provide a unique strategic position or a comparative edge to the organisation. This argument has affinity with the resources based view (RBV) of an organisation, which suggests that organisational resources complement each other and thus constitute a common culture to execute business. RBV asserts that the strategic success of an organisation depends upon productive organisational resources, such as financial, physical, individual, and other organisational attributes, or a collection of these attributes (Penrose 1995).

RBV theory has three aspects which contribute to the competitiveness of the organisation. These three aspects are resources (Wernerfelt 1984); capabilities that allow an organisation to organize and develop these resources (Stalk *et al.*

1992); and competencies that enable an organisation to put into practice corporate strategies (Prahalad and Hamel 1990). Considering these three features, strategic asset planning process takes a RBV approach (figure 2-2) and treats assets and their management as a strategic priority that must be aligned with other priorities such as, production/manufacturing/service provision costs, quality, and order/service provision delivery speed (Davies and Kochhar 2002; Dangayach and Deshmukh 2001; Schroeder *et al.* 1989). The strategic asset planning process addresses asset demand and attempts to fulfil the value profile that the stakeholders and regulatory requirements attach to the use of asset by aligning these priorities with market requirements (Llorens-Montes *et al.* 2004; Kerr and Greenhalgh 1991).

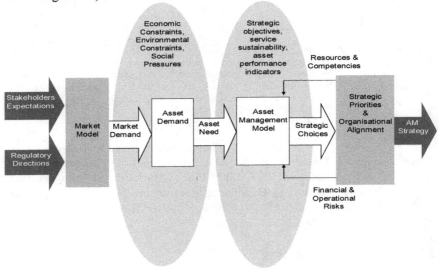

Figure 2-2: Asset Strategy Development Process
(Adopted and modified from UMS 1999)

Asset planning process takes into account the business objectives and thus considers the type of assets needed to attain those objectives. Consequently, the process is aimed at recognising the gaps between the envisaged performance level of current asset solutions and the level required to address market demands. This gap analysis provides the basis for demand management strategies, which are aimed at mitigating the need for new assets. According to IIMM (2006), asset demand is typically driven by one or more of the following factors:

a.   economic constraints, i.e. the affordability of investment,

b. environmental constraints, such as limitations in water resources, deg-
    radation of waterways and air quality or climate change impacts, and
c. social pressures, such as the impact of asset on community values.

Demand management strategies are subjected to economic, technical, social, environmental and sometimes political constraints, and are aimed at reducing peak demand (which is the major reason for an asset solution operating at maximum capacity). Reduction in peak demand, however, is realized by changing the demand to off peak intervals, so as to achieve the balance in asset capacity throughout (for example, the variable electricity pricing to encourage consumers to shift their consumption to off peak times). The other aim of demand management is reduction in average demand, seeking to modify both peak and base flows. This approach is applicable where there are constraints in resources (for example scarce water source), financial gains to be made (for example need to reduce operating costs), or there is an adverse environmental impact to be addressed (for example encouraging alternative means of transport to private car). The objective of demand management, however, is to actively seek to modify customer demands for services in order to:

a. optimise utilisation/performance of existing assets,
b. reduce or defer the need for new assets, and
c. meet the organisation's strategic objectives (including social, environ-
    mental and political), deliver a more sustainable service, and respond to
    customer needs.

Principles of asset management (see section 1.2.2) are based upon the alignment or the strategic fit (Porter 1980) of the organisation's resources to best meet stakeholders' needs. These needs, however, vary with time. Asset need management, therefore, represents an ongoing process to obtain strategic alignment between organisation's objectives and market demands, which drives the asset management model of the organisation. According to Boyd (2001) the strategic choices that an asset management model presents characterize the objectives that an organisation associates with the use of an asset. These choices are often the set of decisions about what the assets can deliver towards business needs, rather than what the assets are designed to deliver. These choices, therefore, require review of the internal operating environment of the asset managing organisation, as well as the community, environmental, financial, legislative, institutional, and regulatory constraints within which it operates. This review provides the directions that shape asset management strategy, by taking into account the strategic priorities, organisational competencies, and how these competencies could best be used towards maturity (financial and non financial) of the organisation. The three most important features of strategic asset planning could be summaries as,

a. providing planning and control of the technical efforts for engineering asset management;
b. providing planning and control of the operational efforts for engineering asset management;
c. enacting a balanced and integrated approach to meet the requirements of asset lifecycle, by addressing areas such as business plan, environmental and occupational safety plans, emergency plan, human resource development plan, asset performance monitoring plan, logistics support plan, and regulatory plan (Eerens 2003).

2.4 Core Asset Management

Core asset management processes are derived from the asset management strategy and are enabled through various operating plans and procedures. These include processes relating to asset design, acquisition, construction, and commissioning; operation; maintenance; refurbishment; decommissioning; and replacement. Core asset management consists of three cycles, i.e. primary asset management cycle, learning and change cycle, and renewal cycle (figure 2-3).

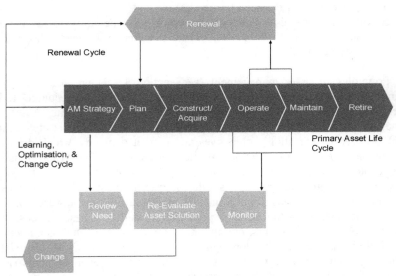

Figure 2-3: Asset Management Cycles

The figure illustrates that primary asset management cycle is derived from the asset management strategy and includes asset construction and commissioning; operation; maintenance; and retirements stages. The figure also explains how learning and optimisation, and renewal cycles are initiated and what impact they have on the primary asset lifecycle. The learning, optimisation, and change cycle is aimed at change, enhancement, and maturity of an asset solution in response to factors such as asset need redefinition, technology refresh, environmental and regulatory concerns, and economic tradeoffs. However, the crucial factor in this cycle is the ability of the organisation to continuously evaluate primary asset lifecycle achievements and compare them with the strategic business objectives. This gap analysis provides learnings on effectiveness of the existing asset solution in meeting the strategic needs of the organisation. The objectives of this exercise are, firstly, to identify enhancements in asset solution design, and secondly (if the first is not possible) to provide alternatives for asset renewal. In doing so, the learning, optimisation, and change cycle calls for redefinition of asset strategy, whereas the renewal cycle informs and necessitates adjustment of asset management plan.

2.4.1 Asset Creation/Acquisition and Supportability Design

The first phase of core asset management deals with the planning, acquisition, construction and commissioning of new assets, as well as design for the supportability of asset lifecycle management. There is strong evidence that manufacturing/production/service provision in terms of cost, quality and timing is influenced by the quality of asset design, choice of technology, and the organisational characteristics (Bertodo 1989; Rao and Gu 1997; Twigg 2002). The key to asset design/acquisition/commissioning project management is a futuristic focus on risk management of design, operation, and maintenance of new or refurbished assets for enduring benefits of the organisation. This risk management consists of detailed assessments by considering the design of the asset, its operational constraints, and maintenance demands.

At the asset design phase organisations are faced with a situation where they have to choose between one of the two available choices. In the first choice, organisations can invest heavily in the design and manufacture/construction of the asset and ensure that the asset is technically sound. This choice, obviously, will result in relatively modest non routine maintenance demands. On the other hand, organisations can opt for a basic asset design and aim to invest heavily in maintenance and refurbishment activities. In this case, organisations aim to take advantage of the continuously evolving technology and are incessantly engaged in updating the as designed capabilities of the asset. There is, however, no right or wrong approach and the choice actually depends on the nature of business and strategic orientation of the organisation. Risk management, therefore, calls for an integrated approach to design, which brings together different functions of the asset management (such as maintenance and operations) to allow for an all inclusive asset and lifecycle support design. The purpose of support design is to provide infrastructure as well as technical support to ensure smooth functioning of an asset from design through to asset retirement. Ettlie (1988) argues six methods for design integration, i.e. constituting design/manufacturing teams; compatible CAD systems; common reporting positions; design for manufacturing; engineering generalists; and research and development in lead time reduction. Trygg (1991), however, identifies two critical aspects that may influence synchronization of design integration, i.e. technological factors, such as material tools and advanced manufacturing technologies; and organisational factors, such as, culture, structure, and people. An appropriate design approach therefore represents diversified functional representation through open collaboration and information exchange, with a balanced mix of technology and organisational support. The aim of design integration is to design a reliable asset solution, by taking into consideration the requirements of various asset lifecycle stakeholders and the technical capabilities of asset and lifecycle support infrastructure.

The major focus during design phase is on designing the asset and lifecycle support solutions for reliability, yet little progress has been made in terms of improving asset efficiency through innovations and advancements in design, for example the reduction of maintenance by integrating reliability, maintainability, human factors, supportability, and quality characteristics in asset design (Blanchard 1997). Bennett and Jenney (1980) conducted a study in reliability aspects of asset designs and concluded several factors that impede reliability provision through design. Although, the study is historic, yet these factors are still prevalent in contemporary asset management paradigm. These factors include, increased dependence of assets on automation technologies, thereby exposing them to new vulnerabilities; limited computerization and recording of information on aspects that may affect design reliability; and high emphasis and high costs of maintenance. The authors further noted that that reliability approach also requires detailed dialogue with asset manufacturers, so as to understand the potential and limitation of as designed characteristics of asset; and detailed data collection of factors impacting reliability of and their analysis at the design stage. In another study, Jarvinen et al. (1996) investigated 44 asset managing organisations and concluded that more than one third of the asset breakdowns occurred due to design based flaws. Lycke (2000) further elaborates on this claim and contends that when new assets are commissioned, problems are frequently encountered during testing, implementation, and turn-on, even though asset design, construction, and installation do not detect any issues. Therefore, it is important to consider design effectiveness by analysing asset solution for reliability and maintainability in order to meet the needs of the overall asset management (Bellgran 1998; Almgren 1999; Jonsson 1999). Figure 2-4 presents an integrated design methodology that consists of a series of activities starting with creation of conceptual design derived from asset need statement. The conceptual design specifies the asset operational characteristics, value profiles, implementation, effectiveness criteria, and maintenance and environmental constraints. Effectiveness criteria, however, constitutes cost benefit analysis of asset availability, reliability, maintainability, and supportability of assets. Up to 80% of the production/manufacturing/service provision quality and costs are entrusted by the conclusion of the conceptual design phase (Miles and Swift 1998). Nevertheless, operational requirements, and technical aspects of asset configuration and design determine preliminary asset design

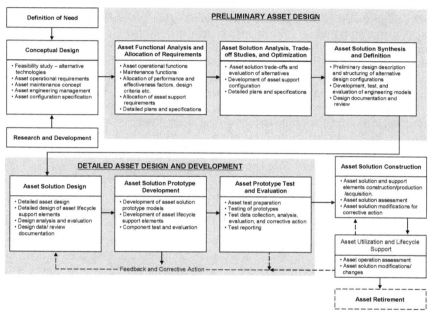

Figure 2-4: Steps in Asset Design and Development
Source (Adopted and modified from Blanchard 1998)

In the preliminary asset design, the first step is to identify the cost efficiency based qualitative and quantitative requirements of asset operation, maintenance, and support. This cost analysis serves as the major constraint or boundary within which the asset solution has to be designed. Within this boundary, alternative asset solution designs are evaluated against various tradeoffs, and the most suitable approach is decided upon. However, supportability analyses for each alternative have a critical and important role in this choice. These analyses determine the resources required to support each asset solution during its lifecycle. Therefore, an appropriate asset configuration and support strategy is identified after having considered all the options. The chosen asset solution's configuration is reassessed for expected overall effectiveness and compliance with the initially developed qualitative and quantitative requirements, in order to cost effectively satisfy the statement of asset need. This then provides the basis for detailed asset design.

During detailed design process design engineers are concerned with how to induce maximum efficiency into asset design; selection of equipment to commission/construct the asset and possible suppliers; preparation of reliability and maintainability predictions; formal and informal design reviews; and test and evaluation of engineering models and prototype equipment. Ylipaa (2000) warns that design based on operational environments (for example, workload and capacity constraints) alone may be insufficient if non-operational aspects (for example, asset health assessment, availability of maintenance and spares) are not taken into account. The author stresses that detailed supportability analysis must be carried out, since they enable the organisation to identify lifecycle support requirements in terms of software, hardware, test and support equipment, spare/repair parts, personnel and skills, training requirements, technical data, facilities, and transportation requirements. Supportability analysis provide for assessment of asset design in terms of cost effective acquisition of support infrastructure. This is followed by the actual construction and commissioning of the asset and phased into full scale operational use.

2.4.1.1 Asset Lifecycle Supportability Modelling

Supportability modelling, as mentioned in the previous section, constitutes a critical step in ensuring reliability of an asset design through its lifecycle. The aim of conducting these supportability analyses is to allow a common baseline for asset lifecycle management; such that all activities can be traced back to the top level requirements (figure 2-5). Asset managing organisations require their assets to operate at a consistent level with regard to capability, capacity, quality, reliability, and costs over their lifecycle (Smith and Knezevic 1996). These organisations take numerous steps to make their assets more reliable and easier to maintain, yet they need more advanced support services than before due to factors such as increasing complexity and integration of asset hardware, condition sensors, and controls (Markeset and Kumar 2005). Even though an asset may be fully functional at the start of its operational life; its operation will result in certain irreversible changes, such as those originating from corrosion, abrasion, accumulation of deformations, distortion, overheating, and fatigue, diffusion of one material into another (Knezevic 1995).

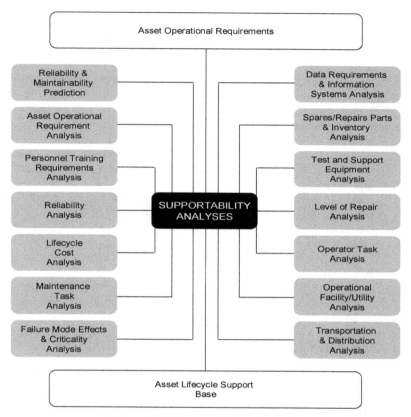

Figure 2-5: Supportability Analysis
Source (Blanchard 1998, p. 170)

These changes on one hand impact the output characteristics of the asset, and on the other hand necessitate new maintenance demands. However, maintenance activities heavily depend on the ready availability of support in terms of spare items, material, trained personnel, tools, equipment, technical and maintenance manuals, and facilities (Knezevic 1992). Considerations of cost effectiveness and efficiency in asset lifecycle management calls for ascertaining the support requirements at the design stage; for example, finding a trade off between logistic support (in the form of spares, personnel, facilities, equipment, tools, etc.), operational availability, and life-cycle cost is an important factor for any asset solution (Kumar and Knezevic 1998). Supportability analyses help in integrating support considerations into system and equipment design; developing support requirements that are related consistently to readiness objectives, to design, and to each other; acquiring the required support; and providing the required support during the operational phase at minimum cost (Galloway 1996). The actual costs and the characteristics of the support required for an asset lifecycle may not be evident until later stages; however, evaluation of support at early stages provides for the commitment of effort, resources, and costs. This commitment shapes the planning for support resources such as repair facilities, personnel, tools, and spares supply and demand management. Blanchard (1998, p128) argues that supportability analysis should ensure that,

a. all facets of asset lifecycle individual stages are covered;
b. all elements of the asset lifecycle are fully recognised and defined; i.e. spares, test and support equipment, personnel and skills, and data requirements;
c. means is provided for relating support requirements to specific functions; i.e. satisfying the requirements of functional design;
d. the proper sequences of activity and design relationships are established along with the critical design interfaces.

Table 2-2 illustrates the description of a selection of these supportability analyses, and highlights the role that they play in asset availability during its lifecycle. It should be noted that the choice of these analyses depends upon the types of assets involved, the level of expertise available in the organisation, level of financial and non financial resources available to the organisation, and the economic tradeoffs made at the time of designing or constructing the asset.

Analysis	Description
Hazard Operability Analysis (HAZOP)	Determination of safety issues by identifying potential hazards, through brainstorming sessions of experienced individuals.
Life Cycle Costing Analysis (LCCA)	Determination of the asset lifecycle and asset lifecycle process cost (design and development, production and/or construction, asset utilization, maintenance and support, and retirement disposal costs); high-cost contributors; cause-and- effect relationships; potential areas of risk; and identification of areas for improvement.
Failure Mode, Effects, and Criticality Analysis (FMECA)	Identification of potential asset failures, the expected modes of failure and causes, failure effects and mechanisms, anticipated frequency, criticality, and the steps required for compensation (i.e., the requirement for redesign and/or the accomplishment of preventive maintenance). An Ishikawa "cause-and-effect" diagram may be used to facilitate the identification of causes, and a Pareto analysis may help in identifying those areas requiring immediate attention.

Continued Next Page

Analysis	Description
Fault-Tree Analysis (FTA)	A deductive approach involving the graphical enumeration and analysis of different ways in which a particular asset failure can occur, and the probability of its occurrence. A separate fault tree may be developed for every critical failure mode, or undesired top-level event. Attention is focused on this top-level event and the first-tier causes associated with it. Each of these causes is next investigated for its causes, and so on. The FTA is narrower in focus than the FMECA and does not require as much input data.
Maintenance Task Analysis (MTA)	Evaluation of those maintenance functions that are to be allocated to the human. Identification of maintenance functions/tasks in terms of task times and sequences, personnel quantities and skill levels, and supporting resources requirements (i.e., spares/repair parts and associated inventories, tools and test equipment, facilities, transportation and handling requirements, technical data, training, and computer software). Identification of high resource-consumption areas.
Reliability Analysis	Evaluation of the asset, in terms of the life cycle, to determine the best overall program for preventive maintenance. Emphasis is on the establishment of a cost- effective preventive maintenance program based on reliability information derived from the FMECA (i. e., failure modes. effects, frequency, criticality, and compensation through preventive maintenance).
Level-of-Repair Analysis (LORA)	Evaluation of maintenance policies in terms of levels of repair (i.e., should a component be repaired in the event of a failure or discarded, and, given the repair option, should the repair be accomplished at the intermediate level of maintenance, at the supplier's factory, or at some other level?). Decision factors include economic, technical, social, environmental, and political considerations. The emphasis here is based on life cycle cost factors.
Evaluation of Design Alternatives	Evaluation of alternative design configurations using multiple criteria. Weighting factors are established to specify levels of importance.
Information and information systems Analysis	Evaluation of information and information systems to meet the information demands of asset lifecycle aimed at increasing responsiveness and creating improvements in asset management processes.

Table 2-2: Supportability Analysis

Source (Adopted from Gerwin and Kolodny 1992; Goetsch 1996; Boyer *et al.* 1996; Lei *et al.* 1996; Small and Chen, 1997; Blanchard 1997; Blanchard 1998)

A detailed breakup of the costs associated with the asset lifecycle is presented in figure 2-6.

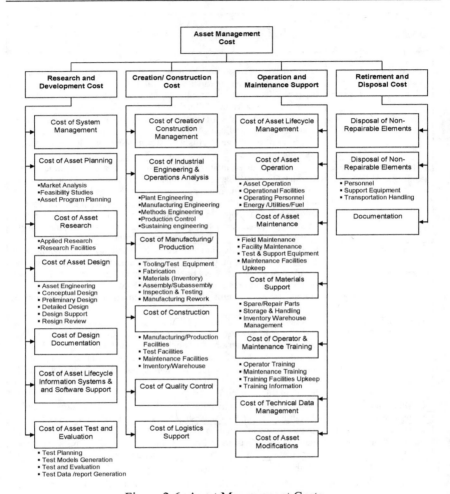

Figure 2-6: Asset Management Costs
Source (Blanchard 1998, p. 84)

This illustration of lifecycle costs highlight two important points, first, that major proportion of the lifecycle costs are centred around the maintenance and support activities; and second, that these costs are attributable to the design and management decisions made during the early stages of conceptual and preliminary design. Nevertheless, the way organisations determines and allocates asset lifecycle costs also reflects their perspective on asset management. Businesses that view maintenance as an investment, manage their assets proactively and their decision making corresponds to a preventative or predictive stance. They prefer to invest in time based periodic maintenance and aim for fixing a failure condition as it develops, thereby aiming to sustain and improve productivity and efficiency of the asset. On the other hand businesses that view maintenance as a cost believe in corrective approaches and operate their assets to failure. This view is further complemented in the survey conducted by Intentia (2004) in table 2-3.

Organisation Size (Employees)	Strongly Agree "Cost"	Generally Agree "Cost"	Indifferent	Generally Agree "Investment"	Strongly Agree "Investment"
1 – 250	12.1%	29.7%	12.6%	27.6%	17.6%
251 – 500	7.0%	25.2%	14.8%	31.3%	21.%
501 – 1000	7.9%	27.0%	9.5%	39.7%	15.9%
1000+	14.5%	23.6%	5.5%	38.2%	18.2%
Total	10.6%	27.5%	11.9%	31.4%	18.4%

Table 2-3: Maintenance an Investment or Cost?
Source (Intentia 2004)

The survey reveals that 49.8% of all survey respondents were of the view that maintenance is an investment. On the whole, large organizations (over 1000 employees) were the most likely to view maintenance as an investment, with 56.4% holding that view. A similar percentage of organizations with 501-1000 employees were of the same opinion (56.4%). Small organizations (under 250 employees) were the most likely to view maintenance as a cost (41.8%). The high percentage of small organizations viewing maintenance as a cost indicates that smaller organizations are more likely to work in breakdown mode rather than in a preventive or predictive mode. Another important indicator revealed in this survey is that large sized businesses follow a proactive approach to asset management; since they have the financial luxury to invest in technologically capable support infrastructure that enables them to be proactive. On the other hand, small to medium sized business opt for corrective maintenance, mainly due to the lack of necessary resources. Nevertheless, important point to note is that these two divergent viewpoints have different expectations of support processes and have different information needs for asset lifecycle decision support. This does not mean that maintenance is necessarily a cost or necessarily an invest-ment. In the final analysis, this depends on the financial position of the organisa-tion and the types of assets.

2.4.2   Reliability Paradigms
Asset managing organisations adopt different reliability paradigms to ensure reliability of their assets. From literature, four reliability paradigms can be identi-

fied, i.e. reliability centred maintenance (RCM), total productive maintenance (TPM), reliability engineering (RE) and control engineering (CE). These methodologies have evolved from different industries and fields of science and are applied to different types of assets.

2.4.2.1 Reliability Centred Maintenance

RCM was developed in the 1960s and 1970s in the aviation industry, with the aim of improving safety and reliability of aircrafts (Jardine 1999). Smith (1993, p. 372), describes the unique features of RCM and argues that it is aimed at preserving asset functions; identification of failure modes that can affect asset functions; prioritising function needs via failure modes; and selecting only applicable and effective preventive maintenance tasks. RCM, therefore, focuses on determining the maintenance requirements of an asset in its operating context, which encompasses entire maintenance strategy formulation process (figure 2-7).

The fundamental aim of activities in RCM is to preserve the asset functions to 'as designed' or 'as constructed' state. Moubray (1997) suggests a seven step approach to RCM. These seven steps represent seven questions that need to be answered in establishing an RCM based maintenance approach. These questions are, what are the functions and associated performance standards of the asset in its present operating context; in what ways the asset may fail to fulfil its functions; what are the reasons behind each of the functional failures; what ensues when a failure happens; how does each failure rate with regard to environment, human safety, productivity losses, and cost; how can a failure be predicted and prevented; and what needs to be done if an appropriate proactive task is not available.

RCM suggests that each asset has at least one function and that there are some requirements on the asset users to maintain the performance standard of the asset, so that it does not reach the failure condition. Consequently, RCM treats an asset as a system within its operating and environmental context, and treats a failure condition as the inability of an asset to do what it is supposed to do. RCM defines a failure condition as a condition that is not acceptable to the user, as opposed to the inability of the asset to perform a task. For example in case of public transport, climate may demand cooling provisions, regulations may demand special lighting, and the remoteness of the destination may demand spares to be carried on board. These may necessary due to the operational context; however, their provision has a direct effect on the performance of the asset and may contribute to a failure condition that may not have been accounted for during asset design. For example, the engine of a vehicle developing maintenance demands due to cooling provisions in the vehicle.

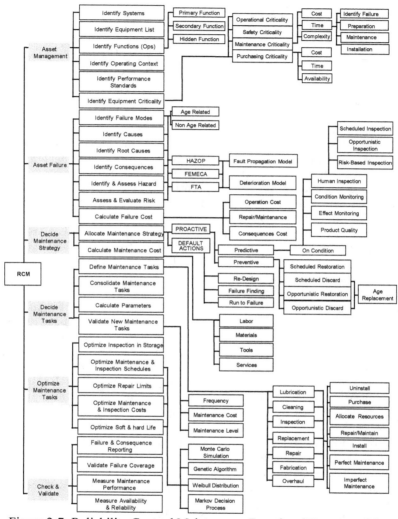

Figure 2-7: Reliability Centred Maintenance Functional Decomposition
Source (Gabbar *et al.* 2003)

RCM distinguishes total failure, where a total loss of functions occur; from partial failures, where there is a loss of a function but the asset is still in running condition (Moubray 1997). Therefore, for each failure the events causing it and the subsequent events are identified for the purposes of predicting the total effects of failure. The failure consequences thus identified are evaluated with regard to human safety, environment, asset operational environment, and costs of maintenance. This evaluation is done to devise a proactive strategy aimed at preventing failure conditions from happening. RCM emphasises human and environmental aspects before material aspects and is rather intense on predicting failure conditions. Therefore, there is considerable timeframe involved in conducting RCM analysis.

2.4.2.2 Total Productive Maintenance

TPM evolved from total quality management and proactive maintenance strategies, with an aim of maximising efficiency of the overall production or manufacturing asset solutions. TPM highlights importance of quality management practices and participation of employees in managing the maintenance function in order to reduce the cost of non conformance to established maintenance schedules and routine. The focus of TPM is the overall asset effectiveness (Nakajima 1988). In order to achieve this, TPM aims to eliminate the six big losses identified by Nakajima (1988), i.e. equipment failure, setup adjustments, idling and minor stoppages, reduced speed, defects in processes, and reduced yield. Figure 2-8 summarises the six major losses an asset may encounter, and their impact on the competitiveness of the business. The origin of these losses can be traced in broad categories of causes, i.e. downtime loses, i.e. loss of asset function due to equipment failure or error in its design, setup, or construction; speed losses, i.e. loss of asset function due to reduction in output speed or inability to perform at 'as designed' output levels; and idling or minor stoppages, i.e. defect losses or loss of asset function due to errors in the processes utilising the asset or the inability of the asset to produce the same quality of output. Overall these losses impact quality, efficiency, and availability of asset operation, which eventually has a negative impact on delivery, flexibility, cost, and quality of the service enabled by the overall business. TPM treats breakdowns under two categories, i.e. function loss breakdowns, and function reduction breakdown. The function loss breakdowns are the conditions in which an asset losses one or more of its functions (Shirose and Goto 1989); whereas function reduction breakdowns are the conditions in which although the asset still operates but it does not provide the same output in terms of speed. It may be noted that these descriptions are quite similar to the total and partial failures in RCM.

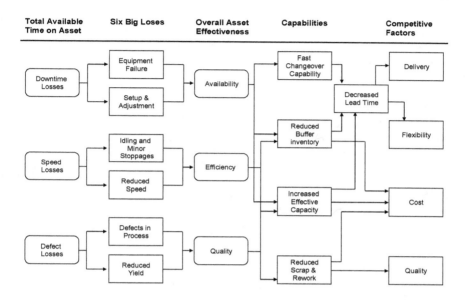

Figure 2-8: Impacts of Disturbances on Competitiveness
Source (Fredendall *et al.* 1997)

Nakajima (1988) distinguishes unexpected, failures and chronic failures or developing failures. Unexpected failures are failures that occur at once and may result in breakdown, and are relatively easy to detect due to the obvious variations in production or manufacturing capacity. In contrast the developing failures have an irregular pattern and are difficult to locate. Furthermore, these failures may occur as a result of physical environment around the asset, for example, dirt, moisture, or temperature variations causing the asset to not to perform at the desired level of output.

TPM calls for standardising asset operating conditions, which requires periodic inspections, cleaning, and servicing. The main objective of TPM is asset efficiency through personnel participation in continuous improvement for preventing sudden and chronic failures. The preventive maintenance preparation takes into account the legal conventions, maintenance standards, asset health history, and maintenance execution history. This, in turn, enhances the organisational ability to identify failures in their developing stages, rather than detecting the failure at a stage when the asset is near total breakdown or shutdown.

2.4.2.3 Reliability Engineering

Reliability engineering, as a reliability paradigm originated from the growth in military electronics after World War 2 (Leveson 2005). It works on the premise that maintenance initiatives are inadequate in correcting the insufficiencies and shortcomings in the intrinsic safety and reliability standards of asset design. It suggests that maintenance only helps in halting the process of degradation and deterioration of these standards. When these standards are not good enough, organisations need to modify the asset design to achieve improved levels of safety and reliability. Reliability engineering, therefore, suggests that design is the most critical aspect in establishing reliability of assets. Reliability engineering aims at enhancing component integrity (by incorporating safety margins for physical components and attempting to achieve error free behaviour of the logical and human components) and uses safety functions during operations to recover from failures (Leveson 2005). These safety functions utilize a range of design approaches to prevent the asset from entering into a potentially hazardous state that may lead to a failure condition.

Reliability engineering, thus, focuses on asset durability though identification of reasons, probabilities, and outcomes of failures, and designs out these errors (failure conditions) with enhancements in component design and materials. In doing so, it relies heavily on statistical analysis, probability theory, and reliability theories, such as reliability prediction, Weibull analysis, reliability testing, and accelerated life testing. Asset equipment and their reliability and failure characteristics are modelled by using mathematical models with reliability networks or fault trees (Vatn 1997), which explains the relationships of elements in an asset configuration and the effect that a particular part or component may have on the operation of the other parts. As opposed to RCM and TPM, reliability engineering is rarely aimed at treating partial failures, mainly due to the difficulties in modelling partial failures.

2.4.2.4 Control Engineering

Control engineering approach is aimed at maintaining the standardised production/manufacturing processes through controlling failures and disturbances in the asset operations (Ibrahim 2006). It suggests that operating parameters of an asset operation, which may be stochastic, should follow a dynamically modelled behaviour. A deviation from these parameters or operating limits and modelled behaviour signifies an atypical state, which specifies a failure condition. In control engineering, dynamic system modelling and system control is considered the science of control and automation engineering. The asset operation variables are controlled with mathematical algorithms, such as PID controllers (Astrom and Hagglund 2000), whereas the controllers maintain satisfactory operations by compensating for disturbances in the process. The term used for undesired asset states is fault (Chiang et al. 2001). Fault detection and diagnosis aim to keep the operators and maintenance informed about the status of the asset operation and to detect the causes of possible faults. Nevertheless, there are three major strategies to detect abnormal operation of an asset, which are data-driven, analytical, and knowledge-based (Chiang *et al.* 2001; Lewin 1995). Data driven models operate purely on measured data and reduce the measured data to lower dimension data without losing essential information of the original asset operations parameters. Analytical methods use mathematical models and parameter estimation to detect faults. Knowledge-based methods use causal analysis, expert systems or pattern recognition to detect faults. The difficulty of the control engineering approach is that the industrial manufacturing and production processes are not always stable. For example, when there is a change in the production plan the system may be in a pre programmed state causing difficulties in control and issuing wrong alerts.

2.4.2.5 Comparison of Reliability Paradigms

In contemporary asset management paradigm RCM and TPM are commonly found. RCM is forward looking and based on futuristic planning aimed at preserving asset functionality, whereas TPM calls for continuous improvements through employee involvement and is aimed at preserving asset efficiency. Reliability engineering is aimed at improving the design, whereas control engineering's objective is control of asset operation and compensation of disturbances by mathematical algorithms. Table 2-4 provides a comparison of all these paradigms and reveals that although all these paradigms are aimed at asset management, yet they differ in their scope and objectives.

	RCM	TPM	Reliability Engineering	Control Engineering
Scope	Asset functionality	Asset efficiency	Asset durability	Asset controllability
Objective	Keeping the asset functionality at the required level	Maximising the asset capacity by equipment efficiency	Enhancing the asset life-time and reliability	Maintaining the production process state
Failure Condition	Inability to fulfil user-required functional capability	Loss or reduction of a capability with regard to optimal performance	Loss of function	Statistically abnormal process state
Lifecycle Stage of Focus	At the asset design and operation phase	At the asset operation phase	At the asset design phase	At the asset operation phase

Continued Next Page

	RCM	TPM	Reliability Engineering	Control Engineering
Overall Focus	Individual assets, users, and manufacturing/ production facility	Individual assets users, and manufacturing/ production facility	Multiple assets, users, and production facility	Single production process
Core Doctrine	Proactive maintenance by preventing failures before they first occur	Personnel participation in continuous improvement for preventing sudden and chronic failures	Design-out failures with enhanced component design and materials	Control of process states and compensation of disturbances by mathematical algorithms.

Table 2-4: Reliability Paradigms
Source (Adopted from Honkanen 2004)

2.4.3   Asset Operation

Manufacturing and production of products and services is subjected to intense changes in technology and varying market demands (Beach *et al.* 2000). The resulting effect on industries is one of strong competition that dictates a shift towards renewal of products and services at regular intervals. This shift is forcing businesses to innovate and update their offerings with added value and features. Shortened product lifecycles and continuous updating of products demands enhanced and reliable asset operation, which calls for proactively assessing asset condition in order to predict developing failure conditions, so that appropriate follow ups could be planned. The essential aim here is to minimise the asset downtime. Therefore, it is important to manage disturbances to increase the overall availability and efficiency of the asset. A disturbance is defined as an unwanted unplanned state or function (Kuivanen 1996) of an asset, which results in a productivity loss.

Apart from technical issues arising out of deficiencies and problems with asset design and technical wear and tear, there are various disturbances that occur due to coordination and management issues with business processes. Research shows that 50% to 60% of the production or manufacturing time is utilized for manufacturing, whereas organisations spend the remaining time in handling disturbances (Ericsson 1997; Ingemansson and Bolmsjo 2004). Disturbances due to process issues generally occur at the supply and customer interfaces of a production or manufacturing system, though can also occur from within the internal business environment. Matson and McFarlane (1999) summarise these disturbances and conclude that disturbances can be,

a.   Upstream disturbances, arising from materials quality and delivery problems, supplier production problems, material property variations.

b. Internal disturbances, arising from information, control and decision making, control and communication system failures, operator errors and omissions, recording/communication errors, materials and stock control problems.

c. Production equipment and labour disturbances, arising from, asset breakdowns, variability in asset performance (quality, cost, production rate), unavailability of labour.

d. Downstream disturbances, arising from make to order issues, such as changes to orders, and quantity and mix variations; and make to stock issues, such as demand variations, forecasting errors, and poor stock monitoring,

Generally these types of disturbances are not directly taken into account in asset lifecycle planning and management. The focus of lifecycle management during asset operation is to keep the asset operating as close to original specifications as possible. It requires continuous monitoring of the asset and asset's operational environment in order to ascertain the ability of asset to meet required service levels and mitigate operational risks. Condition monitoring of an asset and its operation is carried out through a series of manual and automated inspections and assessments. The interpretation of resulting information indicates any deviations from normal asset operation, in which case it specifies nature and timing of the follow up actions to increase the availability and reliability of asset operation.

2.4.4   Asset Maintenance

Traditionally, asset maintenance function has been the major focus of research and industry focus from among all the activities in asset lifecycle management. Maintenance strategies include a variety of tasks and actions aimed at increasing reliability of assets, and are derived from the reliability paradigms (see section 2.2.3) that the organisation conforms to. However, key decision elements that govern maintenance regime are cost effective mix of planned and unplanned maintenance to minimise the total maintenance costs (figure 2-9). Total costs of maintenance increase with the operational age of the assets. However, reactive maintenance costs decrease with the operational age of the assets, whereas the proactive maintenance costs increase with the age of the asset.

Planned maintenance represents activities such as servicing, routine corrective maintenance, planned preventive/predictive maintenance, and planned disposal; whereas unplanned maintenance represents emergency corrective maintenance, modifications, and redesigns arising from sudden wear out (IIMM 2006). Moubray (1997) divides failure modes into two categories, i.e. hidden failures, where the asset operator doesn't become aware of the loss of asset functionality, under normal circumstances (for example, chemical additives in oil affecting its viscosity, which consequently affects the output of motor); and evident failures, where the asset operator becomes aware of the fault due to an evident break down.

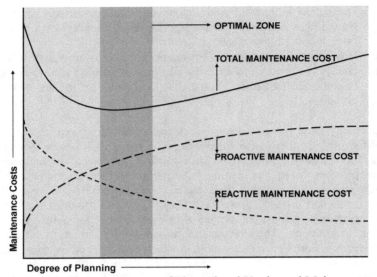

Figure 2-9: Financial Impact of Planned and Unplanned Maintenance
Source (IIMM 2006, p. 3.95)

The foundation of maintenance is the ability to identify possible failure modes. A failure mode, however, is the way an asset fails to perform its usual operation or function. Failure is reached in one of the three ways, as defined by El-Haram (1995), i.e. when an asset become inoperable suddenly and can no longer perform its required operations; when an asset cannot fulfil some or all of its operations at the same performance standard as originally specified; or when an asset gradually deteriorates to an unsatisfactory level of performance or condition, and its continued operation is unsafe, uneconomical or aesthetically unacceptable.

Tsang (2002) argues that management's considerations about maintenance deal with four strategic dimensions, i.e.,

a. service delivery options, which characterize the choices between in-house and outsourced service;
b. organisation and structuring of maintenance functions;
c. selection of maintenance methodology; and
d. design of maintenance infrastructure and supports.

Maintenance approaches are aimed at fixing failures, and are classified into two categories, i.e. reactive maintenance, and proactive maintenance. However, there are three major approaches to maintenance (Kelly 1989; Davies 1990; Gits 1994; Niebel 1996; Moubray 1997), i.e. failure driven maintenance; time based maintenance; and condition based maintenance.

Failure driven maintenance, also termed as corrective maintenance (Moubray 1997) is a reactive approach that calls for running an asset to failure. In this approach, elementary maintenance consideration is given to maintain the health of an asset during its operation. Only basic servicing is carried out and the asset is operated till it reaches a failure condition (Tsang 2002). Choice of failure driven maintenance may be motivated by cost, maintainability constraints, technical obsolesce, or end of need considerations; and is dominated by unplanned occurrences, such as malfunctions, functional failures, or total asset breakdown (Niebel 1996; Tsang 1995). In contrast, time based maintenance is a proactive approach and is also termed as periodic preventive maintenance. It is carried out as a routine activity with the underlying belief that routine inspections and servicing options reduce the severity and frequency of failure occurrence (Percy and Kobbacy 1996).

Condition based maintenance (CBM), which is also known as predictive maintenance, aims at carrying out corrective maintenance when an asset or component has reached a pre-determined condition before failure or breakdown (Tsang 1995). This maintenance strategy is in direct contrast to the periodic preventive maintenance, where an asset undergoes routine maintenance. Preventive maintenance is often criticized for wastage of effort and resources, and not accounting for the maintenance issues that develop between scheduled periods of maintenance. On the contrary, condition based maintenance is aimed at continuous monitoring of asset condition so as to predict failure condition in its development stage. It depends on condition information such as, vibration measurement and analysis; infrared thermography; oil analysis and tribology; ultrasonics; and motor current analysis. For example, vibration analyses reveal wear, imbalance, misalignment, loosened assemblies or turbulence in a plant with rotational parts (Tsang 2002).

The nature and detail of work carried out in maintenance depends upon the criticality and complexity of the asset. However, successful maintenance work completion requires effective project management, which takes into account factors such as availability of information on failure root causes, availability of spares and maintenance expertise, location of workforce, structuring of maintenance work, integration of maintenance management systems with other management information systems, and management of maintenance interface with asset operations (Tsang 2002). Effectiveness of maintenance is measured in terms of asset performance, such as availability, reliability and overall asset effectiveness; cost of asset performance, such as labour and material costs of maintenance; and process performance, such as ratio of planned and unplanned work, and asset operations schedule compliance (Campbell 1995).

2.4.5   Asset Renewal

Asset renewal/restoration/rehabilitation is aimed at revamping the asset solution to ensure that the required service levels can be maintained or attained. This situation occurs when it is too costly to maintain an asset so that it keeps delivering at the original level of service, or there is better technology available. Asset renewal requires considering various failure modes of the asset under question, likelihood of their occurrence, and impact of failure in terms of risks that it poses to the business. For example, if the failure is related to asset capacity, then the operational efficiency of the asset involved should be examined with the view of minimising this risk as part of the rehabilitation/replacement process. Asset renewal process is information intensive and requires the asset managing organisation to take into consideration various options available in terms of full asset refurbishment, expansion, or substitution. It follows an articulated evaluation process to arrive at an information based quality decision. According to IIMM (2006, p2.37-2.38), this decision needs to take into consideration a number of issues, such as,

    a.   the cost of rehabilitation versus replacement;
    b.   the possible increases in effective life following different treatment options;
    c.   probability and consequences of failure if rehabilitation/renewal does not take place;
    d.   benefits that the customers derive from the different levels of service that each option offers;
    e.   capital investment requirements and options;
    f.   future annual and periodic maintenance and operating costs following rehabilitation or replacement; and
    g.   justifications for any premium being paid for increased level of service.

2.4.6   Asset Disposal/Retirement

Asset retirement/disposal is an activity that has not received due attention in research and industry, although in certain cases it has raised various concerns about environmental degradation. Asset disposal/retirement represents investments aimed at replacing or deconstructing an existing asset through end of need; technology refresh; or under utilisation due to as-designed limitations, condition beyond repair, or decreased demand. This occurs when operating and managing the asset is financially not feasible for the organisation. The other needs for asset disposal may arise due to the organisational consideration of enhanced level of service, and/or to improve asset configuration, and/or change in physical location of the asset that warrants disposing off of the existing assets and constructing, acquiring, or setting up new ones. Asset disposal aims at sustainable rationalisation of an asset according to safety/regulatory/environmental requirements when it is no longer in service. According to IIMM (2006, p2.38) decision on asset disposal requires weighing some options, such as

a. adequacy of reasons of disposal;
b. availability of more effective and efficient way of providing service in future;
c. legal/environmental/social/heritage barriers to disposal;
d. availability of disposal options; residual value of the asset(s);
e. physical disposal problems; breaking down and selling of asset components, or reusing the asset components in similar assets;
f. utilisation level of the asset to be disposed; ways of offsetting asset replacement/disposal; and
g. asset technology obsolesce.

New South Wales Government asset management committee proposes six stage asset disposal planning process (NSWG 2001). The process begins with identifying the assets that have become surplus to the service delivery, followed by the comparison of benefits of retention and disposal. This comparison provides a value maximisation statement, which specifies the disposal mechanism. Finally, a detailed disposal plan is prepared and implemented. This plan defines the costs involved in disposal and techniques for sustainable physical disposal of assets.

2.5   Asset Lifecycle Performance Assessment

Performance evaluation of the asset lifecycle management plans, processes, and enabling technologies provides the progress report on effectiveness of existing initiatives and highlights the underperforming areas or gaps in performance. Evaluation exercise thereby enables a continuous improvement regime for asset solution as well as lifecycle support. Asset lifecycle performance assessment is not necessarily an internal business activity, as regulatory agencies may also carry out such exercises. The objectives of an external audit may be to examine asset operation for compliance with environmental and legal regulation, whereas internal business evaluations may include these objectives as well as for the measurements of existing asset operation and support infrastructure efficiency, effectiveness, and reliability against the desired and planned levels of asset throughput. Furthermore, performance evaluation may be aimed at different dimensions of asset lifecycle management, such as, effectiveness, reliability, and cost effectiveness of design, operation, and maintenance. Nevertheless, according to IIMM (2006), asset lifecycle performance evaluation is aimed at,

a. evaluation and justifications of planned levels of service;
b. compliance with monitoring, and reporting and requirements;
c. compliance with the planned techniques and methodologies to enable cost effective asset lifecycle treatment options, such as risk management, predictive modelling, and optimised decision support;
d. evaluation and identification of task priorities and resources requirements;
e. evaluation and justification of the roles and responsibilities for various organisation units in relation to asset management activities;
f. evaluation of information requirements of asset lifecycle; and
g. continuous improvement of asset management plan.

The choice of evaluation methodology differs from business to business. There are a variety of evaluation models, frameworks, and techniques available that aim to evaluate different aspects of an asset lifecycle. Some of these are qualitative while others are quantities, yet there are some that embody both characteristics. However, review of literature reveals that a good performance measurement system should include measures that aim at:

a. informing business strategy, in compliance with the critical success factors, customer requirements; and include evaluation of financial as well non financial aspects (Clarke 1995; Manoochehri 1999; Bititci 2000; Bititci *et al.* 2005);
b. meeting the requirements of specific situations in relevant manufacturing operations, and be easy to understand and implement (Santori and Anderson 1987; Ghalayini *et al.* 1997);
c. a fast and rigorous response to changes in the organizational environment (Bititici *et al.* 1997; Medori and Steeple 2000).
d. focusing on business processes as well as the structures that deliver value (Neely *et al.* 1996);
e. integrating different aspects of asset management to constitute a chain for business competitiveness (Suwignjo *et al.* 2000);
f. balancing needs of various stakeholders, such as customers, third party service providers, and regulatory agencies (Neely *et al.* 1996);
g. competency development and business intelligence infrastructure development in order to create and sustain value for asset management processes (Neely and Adams 2001);
h. stimulating the continuous improvement processes (Kaplan and Norton 1992; Medori and Steeple 2000); and
i. linking measurement system to the reward systems (Tsang *et al.* 1999).

2.6   Information Flows in Asset Management

Asset lifecycle management is information intensive. The variety of asset lifecycle processes generate, process, and analyse enormous amount of information on daily basis (Hudson *et al.* 1997). Information systems utilised for asset management not only have to provide for the control of lifecycle management tasks, but also have to act as instruments for decision support. Asset lifecycle management can, thus, be viewed as a combination of decisions associated with strategic, tactical, and operational levels of the organisation (figure 2-10).

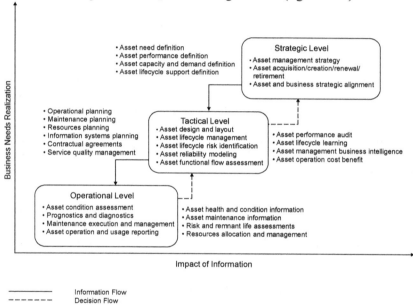

Figure 2-10: Information and Decision Flow in Asset Management
Source (Sardar *et al.* 2006)

It actually represents a two-way street, where decisions flows define what should be done at each stages of lifecycle management, and information flows enforce these decisions and provide feedback on their effectiveness. The operational layer specifies the functions necessary to keep the assets in near original condition. The focus of this layer is to enable asset operation and condition monitoring, diagnostic ability to detect failure conditions, alarm and notifications generation, and maintenance work execution. Information at this layer flows from an array of condition monitoring sensors to operational information systems that utilise this information for a variety of follow up actions and operational level decision support. The objective of this layer is to provide information enabled consolidated view of asset information to tactical and strategic layers. Successful utilisation of information systems at this layer, therefore, depends upon the speedy availability, quality, integration, and interoperability of information.

Tactical layer has a proactive scope and is focused on establishing procedures and plans aimed at developing lifecycle operational, maintenance, and service quality levels; risk management initiatives; and asset and lifecycle processes reliability modelling. These are accomplished through enabling technical and organisational infrastructure. Critical information inputs in this layer are the risk assessments, asset health information, and information about the availability of lifecycle support and maintenance resources. This layer is the most important layer in terms of information systems, as the overall asset management is driven from this layer. This layer specifies the types of information systems employed to facilitate asset lifecycle management and the way they are implemented. Significance of this layer underscores the need to integrate asset management systems and operational technologies like control systems to seamlessly support a variety of lifecycle management tasks such as maintenance scheduling, maintenance workflow management, inventory management, and purchasing; and to integrate these functions with manufacturing scheduling and execution, and risk management. The information thus available facilitates decisions support at the strategic layer. The strategic layer takes a total cost of ownership perspective and is concerned with the decisions such as asset design and refurbishments, asset demand management, performance definition, planning of performance and capacity, and asset lifecycle supportability design. These decisions depend upon the quality and availability of information on asset performance; cost and efficiency audits; and sustainability of the service and delivery potential through asset operation. The operational and tactical layers work towards compliance of performance standards set forth at the strategic layer. Information systems in asset management, thus, act as the bond that brings together or integrates different process stakeholders, functions, and even organisations. In theory, these systems facilitate information enabled view of asset management. However, realisation of such a view of asset lifecycle through information systems requires appropriate hardware and software applications; quality, standardised, and interoperable information; appropriate skill set of employees to process information; and the strategic fit between the asset lifecycle management processes and the information systems.

2.7   Asset Lifecycle Learning Management

Scope of asset management spans engineering as well as business activities; where most of these activities are cross functional or cross enterprise. For example, maintenance processes influence many areas, such as asset operations; asset design/redesign; and safe workplace assurance. The outputs from maintenance are further used to predict asset remnant lifecycle considerations, asset redesign/rehabilitation, and planning for the maintenance support and spares supply chain management. This highlights that the foundation of effective asset management is preserving information relating to asset lifecycle, and synthesising learnings and knowledge from this information, preserving them, and making them available to all functions involved in the management of asset lifecycle.

Asset managing organisations are increasingly implementing information systems to synthesise, preserve, and communicate learnings and knowledge relating to asset lifecycle management. However, results of their implementation have been a mixed one. Most organisations have struggled to make effective use of the information that they collect. This is due to the fact that activities relating to management of assets in these organisations are carried out in isolation from one another, they hardly collect right information, are deficient in functional integration, and lack collaboration to share information and knowledge (Hipkin 2001; Levitan and Redman 1998). Learnings or knowledge sharing, however, requires a cohesive learning environment that supports tacit and explicit exchange of information and rewards such exchanges. It calls for asset management to extend its resource based view to a knowledge based view, thereby building competencies that contribute to competitiveness and responsiveness of the organisation (Curado 2006). Asset lifecycle learnings and knowledge management provides for continuous improvement of asset lifecycle management, whereby the organisation learns from mistakes and makes efforts to not to repeat them. These learnings, thus, allow the organisation to learn, evolve, and mature in terms of internal efficiency and responsiveness to internal as well as external challenges. Lifecycle learnings, thus, provide the focus in an integrated asset management view that is accessible to every function within the asset lifecycle. This integrated view allows planners and decision makers at each lifecycle stage to make informed choices after having considered the likely impact of their decisions on related areas of asset lifecycle.

2.8   Summary

This chapter has examined literature on asset lifecycle management and has highlighted its various stages and related areas. Scope of asset management activities extends from establishment of an asset management policy and identification of service level targets according to the expectation of stakeholder and regulatory/legal requirements, to the daily operation of assets aimed at meeting the desired levels of service. Asset lifecycle management includes asset commissioning, operation, lifecycle reliability assurance, maintenance, decommissioning and replacement. Information systems provide necessary support to enable asset management processes; facilitate decisions support for the overall effectiveness of asset management; and preserve, synthesise, and communicate learnings and knowledge relating to asset lifecycle. However, in order to ensure robust asset management, it is extremely important to evaluate the performance of asset lifecycle management processes and infrastructure supporting and enabling these processes. These evaluations should enable a continuous improvement cycle in the organisation so that these assessments provide actionable learnings, whereby underperforming areas are highlighted and corrective action is taken.

The next chapter builds on the discussion in this chapter and explains the role of information systems in asset management. It discusses the barriers and issues posed to information systems implementation for asset management, and develops the case of information enabled integrated view of asset lifecycle through the application of information systems. In doing so, it also elucidates various perspectives on information systems implementation and introduces a framework for alignment of strategic asset management considerations with information systems.

3 Information Systems for Asset Management – Implementation, Issues and Value Profile

This chapter continues the argument developed in Chapter 2 and applies it to information systems paradigm. It reviews literature to develop understandings of the concepts of information systems implementation in general and information systems implementation for asset management in particular. This chapter argues that information systems are an integral part of organisational evolution and have an active role in its maturity. Organisations interpret, learn and endorse their environment (Daft and Weick 1984), through understanding and magnification of details interlocked in the contexts within which they operate (Weick 1995). It is a continuous process of rotation between particulars (for example information systems) and explanations (for example the variety of roles of information systems), with each giving shape and understanding to the other in each cycle. Thus, the characteristics of the organisation are shaped, as the particulars begin to interact with each other and as the explanations provide for accurate inferences. Therefore, to understand how asset managing organisations deal with information systems implementation, it is essential to understand the roles that these systems have at different levels of the asset lifecycle.

This chapter investigates how information systems facilitate asset lifecycle management by providing for better asset design, operation, maintenance, and retirement/reinvestment. It starts with a detailed discussion of the role of information systems in asset management and the upstream and downstream flows of information associated with each stage of asset lifecycle, followed by how information systems enable an integrated view of asset lifecycle and the issues encountered by engineering organisations with regard to utilisation of these systems. This chapter then highlights various theoretical perspectives on information systems implementation, and the way asset managing organisations should approach information systems alignment with strategic asset management considerations. The chapter, thus, develops the argument that effective information systems implementation for asset lifecycle management depends upon how organisations view technology, their understanding of technology, and the socio-technical context within which these systems are employed.

3.1   IS for Engineering Asset Management

3.1.1   Scope of Information Systems in Asset Management

Engineering enterprises mature technologically along the continuum of stand-alone technologies to integrated systems, and in so doing aim to achieve the maturity of processes enabled by these technologies and the skills associated with their operation (Bever 2000).

Asset managing engineering enterprises have twofold interest in information and related technologies, first that they should provide a broad base of consistent logically organised information concerning asset management processes; and, second that they provide real time updated asset related information to asset life-cycle stakeholders for strategic asset management decision support (Haider and Koronios 2005). This means that the ultimate goal of using information systems for asset management is to create information enabled integrated view of asset management. so that asset managers have complete information about an asset available to them, i.e. starting from their planning through to retirement, including their operational and value profile, maintenance demands and treatment history, health assessments, degradation pattern, and financial requirements to keep them operating at near original specifications. In theory information systems in asset management, therefore, have three major roles; firstly, information systems are utilised in collection, storage, and analysis of information spanning asset life-cycle processes; secondly, information systems provide decision support capabilities through the analytic conclusions arrived at from analysis of data; and thirdly, information systems provide for asset management functional integration. In doing so, information systems for asset management seek to enhance the outputs of asset management processes through a bottom up approach. This approach gathers and processes operational data for individual assets at the foundation level, and on the higher levels provides a consolidated view of entire asset base (figure 3-1).

Theoretically speaking, information systems translate strategic asset management decisions through the planning and management considerations into operational actions. They achieve this by aligning information systems with asset management strategy. The planning and management level defines the design of business processes and choice of technology that enables these processes and aligns operational level with the strategic asset management considerations. Thus, in top down direction the information systems 'translate' strategic asset management considerations into action. On the other hand, from bottom up these information systems provide information analysis and decision support. This decision support allows for assessment of the effectiveness and maturity of existing asset lifecycle processes, enabling technical infrastructure, and management controls.

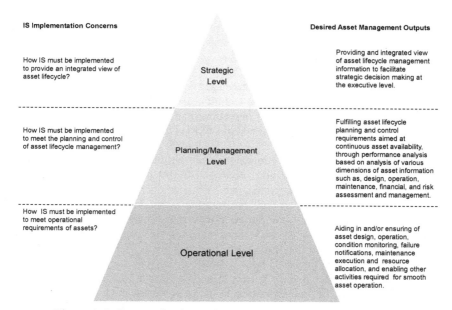

Figure 3-1: Scope of Information Systems for Asset Management

Top management utilises these assessments, at the strategic level, to bridge up gaps in performance or to re-engineer or re-adjust strategic asset management considerations. Therefore, in bottom up direction the information systems act as 'strategic enablers'. In crux, information systems for asset management must allow for horizontal integration of business processes and vertical integration of functional areas associated with managing lifecycle of assets. An important measure of effectiveness of information systems, therefore, is the level of integration that they provide in bringing together different functions of asset lifecycle management, as well as stakeholders, such as business partners, customers, and regulatory agencies like environmental and government organisations.

Information systems at the operational level have to provide for standardised information base that drives the management and strategic levels. In doing so, these systems also have to provide certain level of coupling with business processes. However, it should be noted that loose coupling would not properly satisfy the information needs of business processes, and tight coupling would make the process technology dependent. The minimum requirements from information systems at the operational and planning/management levels are to provide functionality that facilitates,

a.   knowing what/where are the assets the organisation is responsible for;

b.   knowing the condition of the assets;
c.   establishing suitable maintenance, operational and renewal regimes to suit the assets and the level of service required of them by present and future customers;
d.   reviewing maintenance practices;
e.   implementing job/resources management;
f.   improving risk management techniques;
g.   identifying the true cost of operations and maintenance; and
h.   optimising operational procedures (IIMM 2006).

In engineering enterprises strategy is often built around two principles, i.e., competitive concerns and decision concerns (Rudberg 2002). Competitive concerns set manufacturing/production goals, whereas decision concerns deal with the way these goals are to be met. Information systems provide for these concerns through support for value added asset management, in terms of choices such as, selection of assets, their demand management, support infrastructure to ensure smooth asset service provision, and process efficiency. Furthermore, these choices are also concerned with in-house or outsourcing preferences, so as to draw upon expertise of third parties (Dangayach and Deshmukh 2001). The primary expectation from information systems at the strategic level is that of information enabled integrated view of asset lifecycle, so that informed choices could be made in terms of economic tradeoffs and/or alternatives for asset lifecycle in line with asset management goals, objectives, and long term profitability outlook of the organisation. However, according to IIMM (2006), the minimum information requirements at the strategic level are to aid senior management in,

a.   predicting the future capital investments required to minimise failures by determining replacement costs;
b.   assessing the financial viability of the organisation to meet costs through estimated revenue;
c.   predicting capital investments required to prevent asset failure;
d.   predicting the decay, model of failure or reduction in the level of service of assets or their components, and the necessary rehabilitation/ replacement programmes to maintain an acceptable level of service.
e.   assessing ability of organisation to meet costs (renewal, operations, and maintenance) through predicted revenue;
f.   modelling what if scenarios such as, technology change/obsolesce, changing failure rates and risks these pose to the organisation, and alterations to renewal programmes and likely effect on service,
g.   alteration to maintenance programmes and the likely effect on renewal costs; and
h.   impacts of environmental/competitive changes.

3.1.2   Seeking an Integrated View of Asset Lifecycle

Asset lifecycle is information intensive; however information requirements of lifecycle management processes are prone to change due to the continuously changing operational and competitive environment of asset management. The ability of an organisation to understand these changes contributes to its responsiveness to internal and external challenges, as well as its capacity to improve and enhance reliability of asset operations. Thus the real value of information systems for asset management comes from how effectively information systems are mapped to the asset lifecycle management processes, and how effectively they are synchronised with other information systems in the organisation. Bever (2000) summarises the information systems architecture in engineering asset management in figure 3-2.

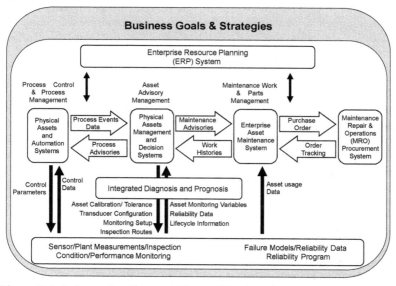

Figure 3-2: Information Systems View of Engineering Asset Management
Source (Bever 2000)

This architecture takes a technical view of information systems for engineering asset management and illustrates the interfaces between different technologies in engineering asset management. However, an information enabled integrated view of asset lifecycle requires integration of asset management core business processes and with these technologies through policies and technical choices to achieve business standardisation, and technical integration and interoperability. Whereas what we have on ground is a technical landscape replete with isolated pools of data that is patchy and error prone; information systems possessing, processing, and communicating data that incomplete and lacks integration; there is a plethora of disparate technology platforms, which make interoperability almost impossible; and to cap it all automation efforts are littered with task technology mismatch (Haider and Koronios 2005).

Chapter 2 highlighted the role of information in asset management, and how its quality, integration, use, and reuse defines the level of success of asset management plans. Information systems and the information residing in them, thus, enable the foundation for planning and execution of asset management processes. This foundation provides for the goal of asset managing organisations, i.e. to use information systems to create information enabled integrated view of asset management. Section 3.1.1 explained this essential goal of information systems utilisation in asset managing organisations and argued that it is a functionally integrated view of asset management related information, starting from their planning through to retirement, including their operational and value profile, maintenance demands and treatment history, health assessments, degradation pattern, and financial requirements to keep them operating at near original specifications. This information enabled integrated view, however, depends upon automation, synchronisation, and integration of planning, execution, control and management of asset lifecycle management activities and processes. As has been argued before the use of information systems is shaped by human action and is dependent upon the socio-technical environment of the organisation. An integrated information enabled view, therefore, depends on the cultural, organisational, and technical maturity of the organisation. The following factors contribute to such an information enabled view (IIMM 2006; Schuman and Brent 2006; Haider *et al.* 2006; Haider and Koronios 2005; Haider and Koronios 2003; Eerens 2003; Woodhouse 2001; Bever 2000; Blanchard 1997)

a.   Asset Design and Configuration Integrity
b.   Asset Capacity and Workload Management
c.   Asset Operational Risk Management
d.   Asset Design Quality Management
e.   Asset Lifecycle Learnings and Optimisation
f.   Asset Need/Value Profile Management
g.   IT/OT Integration
h.   Data Integration
i.   Data Quality
j.   Data Standardisation
k.   Data Interoperability

3.1.2.1 Asset Design and Configuration Integrity

Asset lifecycle management is strongly influenced by asset design specifications. The drawings, diagrams, manuals, reviews, and histories of asset functionality are fundamental to maximising asset effectiveness (Wu 2001). While configuration information provides an understanding of design, design information itself is useful in determining component and process level risk assessments (Burgess *et al.* 2005). Configuration information guides operation, maintenance, and re-design in compliance with the original design, regulatory and operational re-quirements as documented in design specifications, license conditions, and safety analysis assumptions (Wang 1999). In addition, this information is useful in identifying failure modes and the effects of these failures on the overall effi-ciency of the asset lifecycle, as well as operations planning, capacity manage-ment, disturbance management, and health management. In case of asset refur-bishment, the same design specifications are used to ensure that the replacement fits the specifications and meets the operational requirements set forth by design-ers. Availability and quality of design and configuration information, thus, pro-vides the basis for the reliability planning spaning the entire asset lifecycle. Reli-ability planning provides assessments of factors that impact entire lifecycle of an asset; such as identification of lifecycle costs, failure modes, sched-uled/unscheduled asset downtimes, and spares procurement. Structured reliabil-ity analysis spanning lifecycle costing, FMECA, fault tree analysis, failure root cause analysis help in developing reliability driven maintenance plans and feed-back activities to keep assets to their near original condition and configuration (Teng and Ho 1996). However, in order to do so reliability assessments require information regarding planning, design, operations, maintenance, financial out-look of an asset; and use the same for subsequent technical and statistical analy-sis to control the consistency of asset design as well as management and control of engineering changes (Ahmed 1996). This requires integration of a variety of information systems, so that the analysis performed are all inclusive and com-prehensive and include complete information to perform reliability assurance.

3.1.2.2 Asset Capacity and Workload Management
Asset operations are scheduled by keeping in account their design configuration and related reliability assessments. Operational capacity and workload informa-tion of an asset has significance for reliability assurance of asset lifecycle. In-formation on current, planned, and historical manufacturing and production lev-els of assets operation efficiency levels is important for planning asset shut-downs, as well as determining the maintenance demands that these operations may generate. This information is also useful for balancing asset capacity; since overproduction may have consequences for warehouse management and related areas. Therefore, to manage asset capacity and workload the asset managing or-ganisations not only have to integrate design and reliability information with operations planning information, but they also needs to preserve learnings and knowledge relating to asset behaviour (operational, technical) as well as the as-sociated financial and non financial resources.

3.1.2.3 Asset Operational Risk Management

Every asset goes through three phases during its lifecycle, i.e. no defects, initiation and development of defects, and failure (Sherwin and Al-Najjar 1998). Condition of assets is therefore required to be monitored continuously to identify wear and tear patterns (Lee *et al.* 2001). Health assessments require information about the behaviour of the asset, such as vibration analysis, oil analysis, tribology, and electrical circuit analysis; as well as the physical environment within which it operates, such as temperature, moisture, and humidity (Hutton 1996; Haider and Koronios 2004b). In doing so, asset managing organisations aim to detect signs of an imminent failure in its development stages rather than when failure condition has fully developed. This early detection of developing failure and speedy communication of information has significance for quality, availability, capacity, safety, risk, and cost considerations of asset operation and maintenance. Asset designers, operators, and maintainers utilise the health advisories generated by the condition monitoring analysis to eliminate design errors and faults, and to prepare operations and maintenance schedules.

In terms of operational risk management of an asset, the autonomic logistic structure serves as a benchmark. Autonomic logistics structure has been proposed for the F-35 Lightening II Joint Strike Fighter (Figure 3-3), and is a joint venture of US Marine Corps, US Navy, and US Air Force. Logistic structure is aimed at increasing the ground to air ratio of a fighter plane based at a carrier. While in operation, the fighter jet or what the JSF calls an intelligent air vehicle continuously monitors its own condition through the sensors installed throughout the jet. This continuous prognostic is aimed at keeping a record of the components require routine maintenance as well as detecting a failure condition in development; whereby the essential aim is to reduce maintenance time and as a consequence increase operational time. The concept of this prognosis is the same as for any fatal disease among humans, the earlier the prognosis is made the better the chances are to increase the operational life of an asset and its economical operation. Continuous prognostics information is communicated to the base station on the aircraft carrier in real time and based upon these indications the condition of the jet is analysed and maintenance alarms, notifications, and follow up actions are initiated.

These assessments generate recommendations for repairs and parts replacements that determine the type of spares required for carrying out maintenance activities. From this point onwards combat planning and the ground to air ratio of the fighter jet rests with the speed of the supply chain logistics. For management of such maintenance operations there are three important factors to consider. First, it is essential to maintain critical stock levels of essential spares on the ship; second, to order the repair stocks for major overhauls; and third to return the redundant spares to an onshore facility in order to alleviate capacity constraints on board. This supply chain also differs in its characteristics from traditional commercial supply chains. In these circumstances inventory consists of a variety of items; the supply and demands of spares is not as stable as compared to some other supply chains for example for consumer products; and the geographic end points of this supply chain are continuously moving. Conventional supply chain management strategies are therefore not able to cope with the challenges of such operational environments.

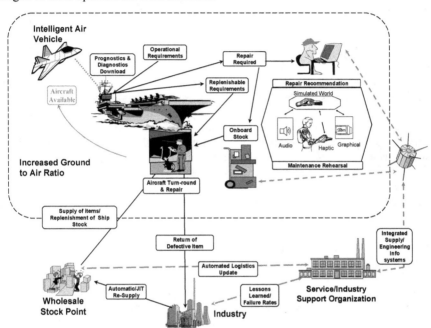

Figure 3-3: Autonomic Logistics Structure – Joint Strike Fighter
Source (JSF 2007)

Asset managing organisations are increasingly turning towards just in time (JIT) delivery of spares. JIT delivery of spares saves costs of holding and agency, as well as logistics costs. However, this also means that a delay in acquiring spares will prolong the asset downtime. Asset managing organisation, therefore, employ maintenance, repair, and operations (MRO) systems to keep information on preferred vendor arrangements and lead time to delivery of replacement parts. MRO along with financial systems provide an integrated view of purchasing, labour utilised in maintenance, inventory acquisition and management, and costs associated with all these factors for management reporting and effective decision support. Nevertheless, the effectiveness of maintenance processes in such stochastic environments depends upon the real time information about the flow of items at each major point in the supply chain, so as to facilitate maintenance planning and resource allocation.

Kochhar *et al.* (1998) argue that engineering organisations are increasingly adopting enterprise resource planning (ERP) systems to provide for integration of information for planning and control functions. The use of ERP systems is aimed at integrating whole of business information, thereby allowing for effective management of resources, staff, materials, and finances (Markus *et al.* 2000). In addition, ERP systems also integrate occupational health and safety information and environment management information with operational information. An appreciation of safety and regulatory issues concerning asset operation and maintenance is critical for decision support on the timelines of performing high risk treatments/repairs. Although the environmental and safety regulatory requirements regarding factors such as pollution controls, hazardous material disposal, noise regulations, may not change too often, their interpretation may vary from site to site and therefore constitutes a important aspect of decision support regarding asset operation and maintenance.

3.1.2.4 Asset Design Quality Management

Asset design quality management largely depends upon the reliability paradigm that the asset managing organisation chooses to support its asset base (see section 2.2.3). These organisations, thus, implement a range of condition monitoring and computerized maintenance management systems (CMMS) to detect failure conditions. CMMS act on the maintenance recommendations generated by the condition monitoring systems, or the planned maintenance programmes to schedule maintenance work execution. Along with planning and scheduling of maintenance work, CMMS assist in allocation of work to maintenance personnel, and acquisition of maintenance equipment. Maintenance work requirements can be quite diverse depending upon the level of treatment required. Reliability analysts review failures, failure modes, and asset operational behaviour on a regular basis to keep the reliability plan up-to-date.

An important aspect of failure review is the root cause analysis that aims to find out reasons for failure and help identify the treatment required and adjustments in the reliability plan to avert future failures. The type of maintenance treatment also specifies recommendations on critical asset lifecycle aspects, such as asset re-design, refurbishment, future operation, asset operation cost benefit analysis, and remanent lifecycle calculations. These recommendations form the basis for economic tradeoffs between asset maintenance and renewal/replacement, effectiveness of reliability regimes, and the emergency response of the organisation.

Asset rehabilitations/renewals are undertaken as a result of technology or need refresh; however asset retirement decisions are taken when asset configuration reaches the end of its intended purpose, when it is too cost intensive to maintain, or when the technology becomes obsolete and is no longer supported by the vendors. These outputs, therefore, provide updates for reliability plan as well as asset management strategic and policy calibrations. The critical factor here is the integration of reliability information with design and maintenance information, and preservation of learnings and knowledge generated from asset's operational behaviour, condition monitoring, health assessments, and maintenance treatments performed on the asset to keep it close to its original design specifications. The availability of these learnings and knowledge to asset designers, operators, maintainers and managers allows them to renew the asset design, better understand the maintenance needs of the asset, and better align asset demand with its operations.

3.1.2.5 Asset Lifecycle Learnings and Optimisation

The information intensive nature of the asset lifecycle suggests that there is enormous amount of explicit and tacit (Nonaka 1995) generated around these processes. This knowledge represents technical knowhow as well as practical knowledge of process execution; competencies required to operate, maintain, and mange the asset; perceptions of organisational systems; and creative abilities (Quinn *et al.* 1996). Learnings gained from asset lifecycle management, therefore, are the combination of information, observations, skills, and experiences used by an individual or a group to enable business process execution and decision making (Sarmento 2005). It, therefore, signifies that knowledge creation, sharing, and management are not automated processes. In fact these processes have social and organisational dimensions that require a culture conducive to learning.

The notion of a favourable culture is particularly important for the transfer and management of tacit knowledge that is possessed by workers. Asset managing workforce, especially the ones involved with operation and maintenance, spend a major portion of their professional lives working closely with the assets. They accumulate and develop valuable knowledge of asset lifecycle, which is lost upon their retirement or changing of jobs. Therefore, capturing, preserving, and managing this knowledge aids in improving efficiency, reducing costs, and mitigating vulnerabilities posed to asset operation (Song 2001). At the same time, it is essential to continuously analyse information contained in the information systems utilised for asset lifecycle management to develop patterns of asset behaviour and financial and non financial profiles of assets. Knowledge management, therefore, becomes a core asset management activity and requires commitment from the top management and staff alike (Bartol and Srivastava 2002) to create an organisational culture that values knowledge sharing.

Researchers have used a number of theories such as, social exchange theory, social cognitive theory, theory of planned behaviour, and theory of reasoned action (see for example Ajzen 1991; Bock and Kim 2002; Lin and Lee 2004; Bock *et al.* 2005) to explain factors that influence knowledge sharing behaviour as well as knowledge sharing culture. Nevertheless, lifecycle knowledge management provides the focal point around which asset lifecycle processes and activities should be organised to create an integrated view of asset management.

3.1.2.6 Asset Need/Value Profile Management

Asset lifecycle learnings provide for assessment of asset solution against its perceived need and value profile. However, when there is an increase in production or service provision demand, it does not always mean that the organisation should deploy more assets. It may not even be possible to deploy new assets, for example, in case of rising electricity demand it may not be possible to construct another hydroelectric dam or thermal power generation plant. In these circumstances asset managing organisations utilise demand management strategies to manage and apportion demand. However, for demand management the availability of information on the value profile of the asset as well as the demand variations is critical.

Organisations that profile each of their asset's financial and non financial behaviour are in a better position to decide on the future usage pattern of their asset base. At the same time, this information is useful for upgrading or refurbishing of asset deign and refocusing of lifecycle reliability support.

3.1.2.7 IT/OT Integration

In the technical domain of engineering enterprises, operational technologies (OT) are as prevalent and important as information technologies. Operational technologies include control as well as management or supervisory systems, such as Supervisory Control and Data Acquisition (SCADA). IT and OT are inextricably intertwined, where OT facilitate running of the assets and are used to ensure system integrity and to meet the technical constraints of the system.

Table 3-1 presents an overview of the characteristics of IT and OT infrastructure. OT set of technologies are primarily used for process control; however, they also include technologies such as sensors and actuators, which are used in many control and data acquisition systems that perform a variety of tasks within the asset lifecycle. Technically, OT is a form of IT as it necessarily deals with information and is controlled by (in most cases) a software. For example, asset operation is continuously monitored for developing failures or failure conditions. There are numerous OT systems employed for condition monitoring at this stage that capture data from sensors and other field devices to diagnostic/prognostic systems; such as SCADA systems, CMMS, and Enterprise Asset Management systems.

Metaphor	Information Technology	Operational Technology
Purpose	Managing information, automate business processes	Managing the assets, technology controlling processes
Architecture	Monolithic, Transactional or batch, RDBMS or text	Event-driven, real-time, embedded software, rule engines
Interfaces	GUI, Web browser, terminal and keyboard	Electro-mechanical, sensors, coded displays
Ownership	CIO, Departmental managers, and knowledge workers	Engineers and technicians
Connectivity	Corporate network, IP-based	Control networks, hardwired
Examples	Finance, accounting, enterprise re-source planning	SCADA, PLCs, modelling, control systems

Table 3-1: IT and OT Profiles
Source (Steenstrup 2007)

These systems further provide inputs to maintenance planning and execution. However, maintenance not only requires effective planning but also requires availability of spares, maintenance expertise, work order generation, and other financial and non-financial supports. This requires integration of technical, administrative, and operational information of asset lifecycle, such that timely, informed, and cost effective choices could be made about maintenance of an asset. For example, a typical water pump station in Australia is located away from major infrastructure and has considerable length of pipe line assets that bring water from the source to the various destinations. The demand for water supply is continuous for twenty four hours a day, seven days a week. Although, the station may have an early warning system installed, maintenance labour at the water stations and along the pipeline is limited and spares inventory is generally not held at water station. Therefore, it is important to continuously monitor asset operation (which in this case constitutes equipment on the water station as well as the pipeline) in order to sense asset failures as soon as possible. However, early fault detection is not of much use if it is not backed up with the ready availability of spares and maintenance expertise. The expectations placed on water station by its stakeholders are not just of continuous availability of operational assets, but also of the efficiency and reliability of support processes. IT or information systems, therefore, need to enable maintenance workflow execution as well as decision support by enabling information manipulation on factors such as, asset failure and wear pattern; maintenance work plan generation; maintenance scheduling and follow up actions; asset shutdown scheduling; maintenance simulation; spares acquisition; testing after servicing/repair treatment; identification of asset design weaknesses; and asset operation cost benefit analysis. An important measure of effectiveness of IT, therefore, is the level of integration that they provide in bringing together different functions of asset lifecycle management, as well as stakeholders, such as business partners, customers, and regulatory agencies like environmental and government organizations.

3.1.2.8 Data Integration

The discussion so far highlights two important points; first, that the intertwined nature of asset lifecycle management processes requires a consistent standard of information; and second, that this information needs to be integrated to enable cross functional processes as well as provides quality analysis for lifecycle decision support. This view is supplemented by Goodhue et al. (1992), who argue that coordination of cross functional processes requires access to consistent data about the processes carried out by different functions. Therefore, the challenge posed to an integrated view of asset lifecycle is of mutual understanding of information contained in information systems employed to support asset lifecycle. However, in contemporary asset managing organisations most of these systems are acquired as commercial off the shelf, which store information in a proprietary format, which makes information exchange difficult (Haider and Koronios 2003). Consequently, simple actions like condition monitoring, and failure notifications are difficult to manage. For example condition measurements are typically collected by disparate information systems, hence all-inclusive assessments cannot be realised and therefore comparisons to reveal asset health status and related lifetime predictions are rarely made. This lack of correlation between operational technologies, enterprise resource planning, process control, maintenance management, and asset health systems impedes quality process enablement and holistic decision support for asset management.

3.1.2.9 Data Quality

A corollary to data integration is data quality. Even integrated data is of no use, if its quality is compromised. Data quality and its integration in business processes has been handled from a variety of angles in literature (see for example, Wang and Strong 1996; Redman 1996; Strong 1997; Naumann and Rolker 2000; Jarke et al. 2003).

Data quality is a broad term and literature reflects its many different definitions; however, Wang et al. (1994), Tayi and Ballou (1998), and Orr (1998) term data quality as 'fitness for use'. The brevity of this definition covers most important aspects of data usage. Orr (1998) argues that the issue of data quality is intertwined with how users actually use data in the system, since the users are the ultimate judges of the quality of the data produced for them. In engineering asset management, the issue of data quality has its roots in multiplicity of data acquisition techniques and methodologies, and the processing of the data thus captured within an assortment of disparate systems. As a result, the information requirements of asset management processes are not properly fulfilled. Therefore, conforming to user requirements is extremely difficult. Non conformance in not the only cause of imperfect data quality, as figure 3-4 reveals the major causes of inadequate data quality in engineering asset management systems. These issues of data quality reveal that data quality has technical, organizational, human, economic, aesthetic, and semiotic dimensions. At the same time, they also reveal that data quality requirements cannot be 'exact' and hence are not presumptively or deductively definable.

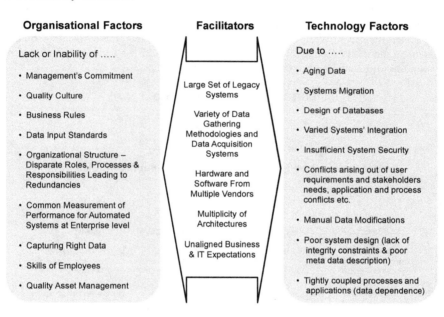

Organisational Factors

Lack or Inability of .....

• Management's Commitment

• Quality Culture

• Business Rules

• Data Input Standards

• Organizational Structure – Disparate Roles, Processes & Responsibilities Leading to Redundancies

• Common Measurement of Performance for Automated Systems at Enterprise level

• Capturing Right Data

• Skills of Employees

• Quality Asset Management

Facilitators

Large Set of Legacy Systems

Variety of Data Gathering Methodologies and Data Acquisition Systems

Hardware and Software From Multiple Vendors

Multiplicity of Architectures

Unaligned Business & IT Expectations

Technology Factors

Due to .....

• Aging Data

• Systems Migration

• Design of Databases

• Varied Systems' Integration

• Insufficient System Security

• Conflicts arising out of user requirements and stakeholders needs, application and process conflicts etc.

• Manual Data Modifications

• Poor system design (lack of integrity constraints & poor meta data description)

• Tightly coupled processes and applications (data dependence)

Figure 3-4: Causes of Dirty Data in engineering asset management
Source (Haider and Koronios 2003)

Data quality problems, however, can roughly be divided into two categories (figure 3-5), such as single source and multi source problems, each of these two categories is further divided into two sections, i.e., schema and instance related problems. Schema level problems may replicate in instances, and can be resolved at the schema level by an improved schema design, schema translation and schema integration. Instance level issues signify data errors and inconsistencies not visible at the schema level. Most common examples of dirty data (Hurwicz 1997; Jarke *et al.* 2003) are,

a.   Format differences.

   b.   Information hidden in free-form text.
   c.   Violation of integrity rules (e.g., in the case of redundancy).
   d.   Missing values.
   e.   Schema differences.

Nevertheless, data quality is relative concept, as its criteria differ from one set of applications and processes to another. What's more, subjective information requirements, which are determined by information user; and objective information requirements, which are determined by information processes, are only theoretically identical. Wang *et al.* (1995) and Helfert (2001) provide data quality measurement and management methodologies that address technical, human, and semantic dimensions of data quality.

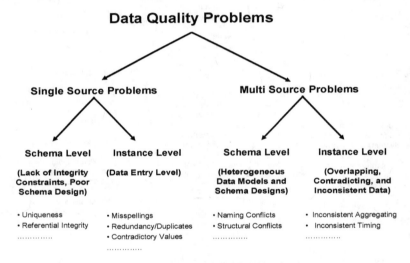

Figure 3-5: Classification of Data Quality Problems
Source (Rahm and Do 2000)

3.1.2.10 Data Standardisation

Standardisation of data refers to consistency of data formats throughout the information systems for asset lifecycle management. This means capturing data in a standardised format, and exchanging and aggregating it with the same level of standardisation. The major advantage of data standardisation is process integration and compatibility. There are a variety of information systems utilised for process automation in asset management, with each supporting different technologies and data standards. The data thus generated cannot be utilised properly and layers of software applications need to be added to the asset management information systems infrastructure to make it consistent with other information. This data can now be used and reused for automation, planning, control, and management of asset management processes. It is, therefore, critical to have all the data to be available in a standardised format for standardisation of processes. Figure 3-6 illustrates an information framework for information systems for asset lifecycle management.

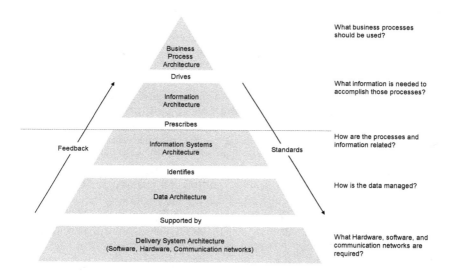

Figure 3-6: Asset Management Data and Business Processes
Source (Adopted from McNurlin and Sprague 2002, p. 66)

The framework suggests that at operational level information systems should capture information in a standardised format, through compatible technology systems, and using standard practices. This standardisation facilitates information exchange between intertwined asset management activities and processes to provide value added results. It, thus, facilitates integration of information and enables quality information exchange for process optimization. Using this framework, information systems implementation for engineering asset management represents a classical top down approach where the process information requirements are interpreted down to the information acquisition level with the appropriate implementation of information systems. Need for standardisation increases as the requirements make their way down to the technical platform; and in the reverse direction the same systems provide feedback on the effectiveness of technical choices in realising the process objectives. The embedded feedback mechanism available in this framework has particular relevance for core asset lifecycle management processes, which are dependent on inputs from other processes. Thus, the information systems for engineering asset management not only provide for standardised quality information but also provide for the control of asset lifecycle processes. For example, design of an asset has a direct impact on its asset operation. Operation, itself, is concerned with minimising the disturbances relating to production or service provision of an asset. At this level, it is important that information systems are capable of providing feedback to maintenance and design functions on asset performance; manufacturing or production process defects; design defects; asset condition; and asset failure notifications.

3.1.2.11 Data Interoperability
Data interoperability is related to the above stated issue of data standardisation. In asset management paradigm, there are a range of different hardware devices. As mentioned earlier, these devices support different technologies and the information thus generated cannot be reused for any purpose other than the one for which the information was generated. Not only that there is total technology dependence in terms of process execution, but the integration of this information also becomes a major problem. Data or information interoperability is, thus, required to ensure smooth flow of asset lifecycle information without any restriction of device, system, or software. Interoperability not only brings together various systems but also contributes to functional and organisational integration. This consistency is vital for the integration of the asset management value chain, which brings together all the different functions of asset lifecycle management and allows for holistic decision support regarding asset operation and allocation of resources to keeping them to their as design condition. Asset managing organisations are, therefore, increasingly turning towards conforming to enterprise architectures or their technology infrastructure.

3.1.2.12 Information Enabled Integrated View of Asset Lifecycle Management
Information systems implementation to enable an integrated view of asset management, requires understanding of the structure of technology, the reasons for choosing particular technology and its implementation approaches, assumptions about the context in which technology is to be used, and previous experiences with technology adoption (Gruber 1991). Information systems employed in engineering asset management have larger than usual span; are generally employed in a distributed environment; and are exceedingly multifaceted in terms of their architecture, the processes that they enable, and the way they are integrated with the social and organisational contexts within which they are employed. Summing up the discussion, a framework to enable an information enabled integrated asset management view is presented in figure 3-7.

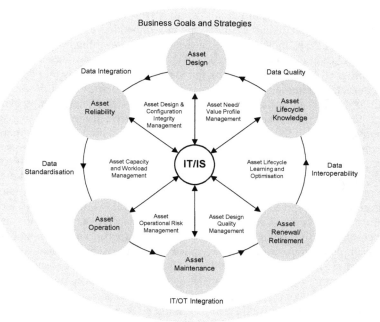

Figure 3-7: Information Enabled Integrated View of EAM

This framework has information systems at its heart, whereby these systems enable asset management processes as well as provide for control of various planning and management activities. In order to fully understand the role of information systems implementation for engineering asset management, consideration must be given to issues like what theoretical and practical assumptions are being associated with information systems implementation; what are the technical, organisational, and social aspects associated with information systems implementation; and how these aspects interact with each other to give shape and meaning to the use and institutionalisation of information systems. The following sections highlight some of the issues resulting from inept implementation of information systems for asset management.

3.2   Barriers to Information Systems Implementation

Value from information systems in the asset management depends upon an assortment of technical as well as organisational and social factors. Effective information systems implementation for engineering asset management, therefore, demands a comprehensive implementation plan that accounts for the aspects that can potentially influence information systems institutionalization in the organisation. Information systems are systems that are embedded in the social structure of the context of their implementation and are, therefore, influenced by the interaction of social and contextual forces. Using information systems signifies the learning progression that is shaped by the view of the technology and the history of information systems management prevalent in the organisation. It characterises the formal and informal organisational structures and relationships evolved over the period of time. The process of interaction between the interacting structures and roles within the cultural context of the organisation shape the maturity of the organisation as well as its technical infrastructure. Information systems, thus, require a certain level of organisational cultural, procedural, and structural maturity to produce enhanced level of service. Organisations need to take stock of this maturity and then select new technologies so that their transition into the organisation is easy and they contribute to the effectiveness of the overall technical infrastructure. It is no surprise, that organisations fail to realize the anticipated benefits of information systems due to lack of appropriate planning regrading their implementation and the way these systems are institutionalised in the organisation (Lederer and Sethi 1996). This research carried out an extensive review of literature to expose the barriers to successful information systems implementation in the context of engineering organisations (see appendix A). The analysis of these barriers reveals some common patterns that highlight the issues and problems that impact successful utilisation of information systems by the asset managing organisation. These issues are discussed in detail in the following sections.

3.2.1   Limited Focus of Information Systems Implementation

Information systems implementation in asset management has narrow focus and scope, which emphasises technical aspects and does not give due attention to organisational, social, and human dimension of technology implementation. This approach to technology implementation at best serves as process automation and does not contribute to the cultural, organisational, and technical maturity of the organisation. There is no attention given to application integration, information interoperability, and data quality. At the broader organisational level, such implementations face resistance from employees and the consequent change management is difficult. As has been argued before that technology is a passive entity and its use is shaped by the interaction of technology with organisational and human factors. Implementation exercises that do not account for the cause and effect relationship that shapes technology are unable to institutionalise technology in the organisation.

3.2.2 Lack of Information and Operational Technologies Nexus

The lack of convergence between IT and OT is a major issue that has technical, management, and organisational dimensions. The root cause of this issue, however, is the fact that IT and OT are managed and owned by different departments within the organisation. IT is generally owned by IT department, whereas OT is owned by the department within which it is deployed. IT is thus managed by IT department and OT is managed by engineers. In the absence of a common set of rules to govern the implementation and use of OT and IT leads to formation of islands of isolated technologies within the organisation, which makes integration and interoperability of technologies cumbersome if not impossible. With limited or no integration, there is poor leverage of learnings and benefits and decision support is unintelligible. Management of IT and OT by different functions results is cost and effort intensive, as this multiplicity of strategies to manage technology (which are essentially of the same stock) cannot connect properly with the business strategy and operational plans. At the same time, this multiplicity also results in the lack of accountability around standardisation of technology and practice, and policies enforcement.

3.2.3 Technology Push as Opposed to Technology Pull

There is an evident lack of commitment from top management in engineering asset managing organisations to institutionalise technology. As a result, IT implementation in general and information systems implementation in particular has been disorganized and is not driven by the strategic business considerations. Most of these technologies have been implemented due to the pressure from regulatory agencies. Thus, these technologies have been pushed into the IT infrastructure of the organisation, without considering the fit between business processes and technology. This lack of user or technology stakeholders' involvement in technology adoption hampers development of a collaborative, creative, and quality conscious organisational culture; and impedes process efficiency. A by product of this efficiency is the inability of the business to collect and disseminate right information that could contribute to organisation wide coordination and horizontal integration. Information systems implementation, thus, is heavily predisposed towards a technology push rather than technology pull strategy.

Engineering enterprises seldom engage in taking stock of their technical infrastructure and the business processes enabled by it (Haider and Koronios 2005). As a result, these organisations are unable to find how well their business processes are performing, how effectively these processes are coupled with technology, and what are the information gaps or requirements that technology has not fulfilled. However, when a technology is selected to fill these gaps, it has a process requirements 'pull' impact and fits in well with the operating logic as well as the enabling technical and non-technical infrastructure of the organisation. On the other hand, when the technology is 'pushed' into the technical infrastructure of the organisation, it has to adapt to the chosen technology. This adaptation has technical, organisational, and human dimensions. As a result there is task technology mismatch and lack of technical standardisation, which gives rise to issues such as information integration and interoperability across the organisation.

3.2.4   Isolated, Unintegrated and Ad-hoc Technical Solutions

Technical infrastructure of an asset managing organisation consists of various off the shelf proprietary, legacy, customised systems and a number of ad hoc solutions in the forms of spreadsheets and databases. Off the shelf systems are developed on customised guidelines and support proprietary data formats; whereas, legacy systems are technologically weak even though they evolve with the organisation. These systems have been in operation for more than twenty years, are developed in old technologies, and are not compatible with new technologies. Ad hoc solutions are developed by employees on their own. They do not conform to any quality or technical standard, and are naturally isolated from the mainstream technology based logical and physical operating model of the organisation. As a result of these anomalies, asset lifecycle information is hard to aggregate, lacks interoperability, and has tight coupling with technology. It, therefore, cannot be reused. Information systems in asset managing organisations are nothing more than isolated pools of data that may serve the needs of individual departments, but do not contribute towards an integrated information enabled view of asset lifecycle management. This means that the existing technical infrastructure in general and information systems in particular are generally not aligned with the strategic asset management considerations, does not contribute to functional integration, and does not conform to a unique enterprise information model.

3.2.5   Lack of Strategic View of Information Systems Capabilities

Information systems implementation in asset managing organisations does not follow a linear path. It is primarily the cost concerns that drive information systems implementation, rather than an approach that takes into account the existing technological infrastructure, business requirements, available skill base, and operational and strategic value of technology investment. At the same time, there is no ex ante or ex post performance evaluation of information systems, which could inform the organisation of the value profile that they technology enables and the issues associated with their implementation and continued use.

Traditionally, asset managers focus on developing the technical foundation for asset lifecycle management around operational technologies and leave the selection, adoption, and maintenance of information technologies to IT managers. This may be attributed to the propensity of asset managers to view information systems utilisation in general as a secondary or support activity to execute business processes. Their emphasis is more on the substitution of labour through technology utilisation rather than business automation and functional integration aimed at internal efficiency and overall strategic advantage. Since the level of input from asset managers regarding choice of information systems has a narrow focus, these systems do not contribute to the organisation's responsiveness to internal and external challenges. There is, therefore, the need for closer interaction between CIO (Chief Information Officer), CTO (Chief Technology Officer), and CEO (Chief Executive Officer) or the COO (Chief Operating Officer). Such a nexus allows for coherent planning, design, and implementation of an organisation's structure, processes, and technical infrastructure, and maturity of its value chain.

3.2.6   Lack of Risk Mitigation for IT infrastructure

Risk management is fundamental to asset management. Almost all asset managing organisations conform to a risk management strategy, standard, or plan; however, their scope does not include the risks posed by or posed to information systems. Risk mitigation within the IT function or department is limited to securing information systems from unauthorised access, intrusion, and malicious codes like viruses. There is no risk assessment, control, and management in terms of business losses occurring as a result of lack of information availability, quality, and integration. A related issue is the lack of information ownership within asset managing organisations, which leads to inability of the organisation to assign accountability for asset management inefficiencies resulting from wrong, fabricated, compromised, and delayed information.

3.2.7   Information Systems Institutionalisation Issues

The issues discussed here regarding information systems implementation for asset lifecycle management are diverse. These issues have technical, human, and organisational dimensions and significant consequences for business development. Information systems implementation should, therefore, not be treated as support activity in the value chain of asset management. It should be pursued proactively and aim to continuously align technology with the organisational structure and infrastructure, process design, and strategic business considerations, so as to realise the soft as well as hard benefits associated with the use of these systems. Thus when information systems will be physically adopted, and socially and organisationally composed, there will be consensus on what the technology is supposed to accomplish and how it is to be utilized. These systems would then provide a learning platform that facilitates organisational evolution and maturity where they act as business enablers as well as strategic translators.

Institutionalisation of information systems is strong underpinned in the political, economic, and cultural context of the organisations, which bring together individuals and groups with particular interests and interpretations and help them in creating and sustaining information systems as socio-technical systems (Bijker and Law 1992). The relationship between information systems and the context of their implementation has been the focus of many research initiatives, such as affiliation between planning sophistication and information systems success (Sabherwal 1999), expediency of strategic information systems planning (Teo and Ang 2000), differences between information systems capabilities and management perceptions (Kunnathur and Shi 2001), impact of inter organisational behaviour and organizational context on the success of information systems planning (Lee and Pai 2003), and identification of key dimensions of information systems planning and their effectiveness (Grover and Segars 2005).

Information systems implementation planning is an intricate task with complex mix of activities (Newkirk *et al.* 2003). It is a continuous process aimed at harmonising the objectives of information systems, defining strategies to achieve these objectives, and establishing plans to implement these strategies (Teo and King 1997). However, as IT environments in general and information systems applications in particular are growing in their control and complexity, information systems implementation is becoming a specialised task and requires broad organisational representation. This broad representation ensures that all aspects of information systems implementation are covered at the planning stage. Organisations, therefore, formulate cross functional teams comprising business managers, information systems personnel, users, unit managers, financial managers, to create an all encompassing implementation strategy these through effective communication and interaction (Earl 1993).

The issues discussed above range from technical issues to social, managerial, and organisational issues. However, the origin of these issues can be traced back to two factors, i.e. inadequate organisational planning and preparation for technology adoption; and disregard of organisational and social change associated with technology adoption. Therefore, the notion of employing information systems requires more than just the installation of technology. It calls for consideration of organisational, technical, structural, processes, and people dimensions of information systems use and the meaning and values that the stakeholders attach to them (Allen 2000). The following sections build upon this theme and develop the case for information systems implementation in engineering asset management.

3.3   Information Systems Implementation for EAM

3.3.1   Defining Information Systems Implementation

Information systems implementation is defined as "an organizational effort to diffuse and appropriate information technology within a user community" (Kwon and Zmud 1987, p. 231). The user community has some aspirations attached to the use of technology, which characterise the values and interests of various social, political and organizational agents (Bijker and Law 1992; Ihde 2002). Walsham (1993) notes that information systems implementation needs to cover all human and social aspects as well as account for the areas that influence or are influenced by their implementation. Effectiveness of information systems implementation, therefore, is a subjective term and depends on the maturity of the organisation as well as the context within which they are deployed. DeLone and McLean (1992) propose six dimensions that determine the effectiveness of information systems implementation, i.e., systems quality, information quality, information use, user satisfaction, individual impact and organizational impact. This shows that information systems implementation is not a one off endorsement of technology; in fact it is a continuing process of learning aimed at the evolving use of information systems. Thus, implementation of information systems blends working together with learning in the organisation (Tapscott 1995). Effectiveness of information systems implementation, however, is compromised if relevant change management strategies are not put in place (Benjamin and Scott-Morton 1992). Information systems implementation, therefore, can be defined as, a continuous process aimed at organisational learning through alignment between the organization's strategy and application of information systems within the organisation, where the use of these systems is shaped by the organizational context and actors and guided by the value profile that the stakeholders of these systems attach to the implementation.

3.3.2 Perspectives on Information Systems Implementation

In computer science, implementation is considered as an activity that is concerned with installation of the IT system and applications and is focused entirely on the technical aspects of the information systems development process. On the other hand, in information systems paradigm, implementation is a process that deals with how to make use of hardware, software and information to fulfil specific organizational needs (Kappelman and McLean 1994). This perspective of information systems implementation is generally governed by two quite opposing views. In a technology driven view, humans are considered as passive entities, whose behaviour is determined by technology.

It is argued that technology development follows a casual logic between humans and technology, and is independent of its designers and users. This mechanistic view assumes that human behaviour can be predicted, and therefore technology can be developed and produced perfectly with an intended purpose. This view may hold true for control systems such as, microcontrollers which have a determined behaviour; however, this view has inherent limitations for information systems due to its disregard of human and contextual elements. A corollary to this objective view is the managerial assumption that information systems implementation increases productivity and profitability. Consequently, management decisions are governed by the expectations from technology rather than the means that enable technology to deliver the expectations.

The opposing stance to the traditional technical view is much more liberating and takes a critical view of the deterministic approach of the relationship of technology with human, organisational, and social aspects. This view illustrates that technology has an active relationship with humans, in the sense that humans are considered as constructors and shapers of the use of technology. In this approach, technology users are considered as active rather than passive entities, and their social behaviour, interaction, and learning evolves continuously towards improving the overall context of the organisation. This organisational change, as a result of information systems implementation, is not a linear process and represents intertwined multifaceted relations between technology, people, and a variety of opposing forces, which makes the human and organisational behaviour highly unpredictable. This unpredictability is attracting attention of researchers to uncover the relationship between humans and technology to develop human centred technologies (Checkland 1981; Boland and Day 1989; Orlikowski and Robey 1991; Orlikowski and Baroudi 1991; Walsham 1993, 1995).

The computer science and information systems perspectives on technology implementation are quite divergent, where one considers it as structure and the other as process. Considering it as structure, demonstrates that technology determines the business processes; whereas the process view argues that technology alone cannot determine the outcomes of business processes and in fact it is open to an intentional propose.

Schienstock *et al.* (1999) summarises various perceptions on implementation of technology using different metaphors (see Table 3-2). When these metaphors are viewed in the light of the two views described above, the first three metaphors, i.e. tool, automation and control instrument conform to the technical view. The process metaphor matches the emancipatory view; whereas the organisation technology and medium metaphors are debateable and can conform to either view.

Metaphor	Function	Aim
Tool	Support business process	Increase quality, speed up work process, cope with increased complexity
Automation technology	Elimination of human labour	Cost cutting
Control instrument	Monitoring and steering business process	Adjustment to changes, avoiding defects

Continued Next Page

Metaphor	Function	Aim
Organisation technology	Co-ordination of business processes	Transparency, organizational flexibility
Medium	Setting up of technical connections for communication	Quick and intensive exchange of information and knowledge
Process	Improve information system	Continuous learning

Table 3-2: Perceptions on Technology Implementation
(Schienstock *et al.* 1999)

Review of literature on information systems adoption reveals that researchers have attempted to address implementation of these systems from a variety of different perspectives. At the same time, it also reveals that the value profile that organisations attach to information systems implementation spans from simple process automation to providing decision support for strategic competitiveness. An in-depth literature review of information systems implementation and adoption from 2000 to 2007 has been carried out for this research (see Appendix B). This literature review identifies different theoretical perspectives that have originated from diversified fields of knowledge, such as business management, organisational behaviour, computer science, mathematics, engineering, sociology, and cognitive sciences. These theories can be classified into three broad categories, i.e. technology determinism (such as information processing, task technology fit, and agency theory); socio-technical interactions (such as actor network theory, socio-technical theory, and contingency theory), and organisational imperatives (such as strategic competitiveness, resource based view theory, and dynamic capabilities theory).

Technology deterministic theories adopt a mechanistic view of organisations where technology is applied to bring about predicted or desired effects. Socio-technical theories are focused on the interaction of technology with the social and cultural context of the organisation to produce desired results. Organisational imperative theories focus on the relationships between the environment that the business operates in, business strategies and strategic orientation, and the technology management strategies to produce desired results in the organisation. The following sections discuss these perspectives in detail and examine their role in effective implementation of information systems for engineering asset management.

3.3.3 Technology Determinism

Technology determinism theories are technology centred, where organisational or societal change is enabled by technology adoption. Technology determinists believe that technology is the prime enabler of change and, therefore, is the fundamental condition that is essential to shape the structure or form of an organization. Technology determinism is also referred to as technology push, where the organisation lets technology determine a solution rather than business need driving the solution. It argues that social and cultural shaping of an organisation is characterised by technology, and has minimum or no influence from human and social aspects. Karl Marx is often cited as one of the earliest technology determinists, with his dictums like 'windmill gives society with the feudal lord: the steam-mill with the industrial capitalist' (Marx 1847). This vision takes a utopian view of technology and advocates the intrinsic goodness of technology to organisations and society at large.

Bijker (1995) argues that technology determinism embodies two subtly different principles. The first principle states that technology development follows a progressive path, one in which older technology is replaced with new technology and denying this progression is to intervene in the natural order. The second principle has been purported by Heilbroner (1994), who argues that technologies act on social interactions in a predictable way. In the light of this principle, technology determinism calls for implementation of technology to enable foreseeable changes in business processes, organisational structure, information flows, communication patterns, and functional relationships. It conforms to a checklist approach and stresses that if certain steps are followed, relevant benefits from investments in information systems can be achieved. These steps include development of technology platforms as well as the activities that must be carried out to use them effectively, such as user training, networking, and data management (Agarwal and Sambamurthy 2002; Laudon and Laudon 1998). These initiatives applied as if they are independent of the context and are valid under any condition or circumstance. User training is one such example, where it is often believed that training on different aspects of software or a system enables the users to handle any issue relating to their operation. Whereas humans have varying levels of comprehension, understanding, and expertise, and, therefore, such as belief does not account for social and human dimensions.

In crux, to provide value from information systems implementation, technology determinism disregards organisational, cultural, and social aspects (that may influence or are influenced by technology adoption) even though they are inherently interlinked (Walsham 2001). This approach, however, recognises that technology provides the necessary support to enable business processes in the organisation. Technology implementation and adoption, thus, becomes a linear process that organisations have to go through in order to exploit the full information systems potential.

In this approach information systems implementation is considered as a smooth process due to assumed objectives with an apolitical vision of the organisation, and organisational harmony and stability. In terms of Boulding's theory

(1956) of hierarchy of systems, technology determinism matches control systems, which are governed by predefined targets such as those in thermostats or robots. Similarly, deterministic implementation of information systems is lead by critical success factors and performance indicators embodied in the information systems implementation plan. It is aimed at business automation rather than enabling business strategy, mainly due to the way it disregards humans and other organisational aspects. In these circumstances, the underlying assumption is the predictability of human behaviour, which implies that whole organisations can be structured to accommodate and make use the information systems in specific and predetermined ways. Technology, with its deterministic behaviour, thus creates new principles and standards for business operations that compel organizations to challenge the status quos and find solutions to questions such as, what information systems do, why they do what they do, and how information systems accomplish what they do; which in turn makes them consider alternative choices of available technology (Volkow 2003).

Information systems implementation in engineering asset management has generally followed technology determinism approach, where technology is considered first and human and organisational factors are not considered until after the actual implementation of technology. This may be attributed to the propensity of engineering organisations to exhibit mechanistic attitude towards technology, which focuses on automation of processes rather than viewing information systems as strategic enablers of the organisation. This also explains the heavy leaning towards maintenance activities in the overall asset lifecycle management strategies, and viewing asset lifecycle management activities as a necessary cost rather than premium of smooth asset operation. Consequently, the existing backdrop of information systems implementation in engineering asset management represents a fragmented approach aimed at enabling individual processes in functional silos, and fails to enable integration of asset lifecycle management activities and processes.

3.3.4 Socio-technical Alignment

The socio-technical views in information systems implementation have originated from organisational theory (Kraft and Truex 1994), institutional theory (Van Der Blonk 2000), and sociology (Dahlbom and Mathiassen 1993). Socio-technical approach was introduced in information systems as a way of maximising value and success of information systems implementation (Bostrom and Heinen 1977). Since then it has been applied to a variety of aspects of information systems operation (such as task technology fit) in a broad way, chiefly through the research of Enid Mumford (see for example Mumford and Weir 1979; Mumford 1983, Mumford 1996). It stresses the importance of social choices in implementation of technology within a particular context (Mumford 1983), by employing participative techniques (Checkland 1981). Socio-technical theorists regard information systems as social systems that are shaped by people with varying interests, and argue that human, organisational, and social factors have a direct relationship with information systems (Checkland 1981; Orlikowski and Robey 1991; Angell and Smithson 1991). This view focuses on the change that takes place in response to information systems implementation through the interaction of various actors within the organizational context that shape information systems use (Orlikowski 1992; Davies and Mithcel 1994). The underlying assumption of this approach argues that success of technology implementation cannot be predetermined or predefined; it in fact depends upon the way different social and human variables react to technology adoption within the context of the organization. Therefore, it presents information systems implementation as a bottom up approach that provides means to achieving the ends of organisational objectives (Ciborra and Lanzara 1994; Ciborra 1996); as opposed to the view held by technological determinists that information systems implementation is an end to means.

Orlikowski (1992), with the help of structuration theory of Gidden argues the dichotomy of the nature of technology. The author posits that technology, on one hand, conforms to an intended reality through its well-established intrinsic objective features, such as hardware and software logic. On the other hand, technology -is also subjective and organisational reality is emergently constructed through the social interaction of humans with technology. This view is supported by Ciborra (1996), who argues that improvisation is a significant aspect that helps in building organisational reality. This improvisation happens at all levels of the organisation, and reflects the way an organisation adjusts to technology implementation. Organisational change, therefore, becomes a dynamic activity, as the planning and decision making processes aims to make sense out of the continuously changing organisational context. Walsham (1993) suggests that the following theories help in understanding the interaction between context and processes,

a. computers and cognition, which focus on an individual level and build the understanding of technology and its relationship to human action and cognition;

b. phenomenology and hermeneutics, which treat information systems as interpretive entities having significance and meaning from designer and users' perspective;

c. soft systems methodology, which works on the supposition that for organizational intervention to occur, it is necessary to take into account the different contingent (but not universal) interpretations that different individuals and groups hold;

d. critical theory, which focuses individual emancipation by developing methodologies that promote open communication and explicitly recognize the existence of structures of power and control in organizations; and

e.   post-modernism, which concentrates on the closeness of events, importance of contingent conditions, and challenging future vision of progress.

Working up from bottom, the socio-technical approach focuses on the effects of technology implementation. It focuses on the way technology enabled processes are managed at the operational level. This requires line managers to be aware of the information needs of business processes; capabilities of technologies to enable these processes; skills of employees to operate these technologies; and social, organisational, and cultural context within which technology is implemented. Here the manager deals with a number of uncertainties about technology, organisational evolution and maturity, and culture. For example, even if the relationships between technology and the context is well established and tested in different organisational settings, yet the emergent and unpredictable nature of human action may change the development, requisition, and institutionalisation of technology. This quagmire has been termed as "soft-line" determinism (DeSanctis and Poole 1994). From this point of view, information systems are an instrument of sense making, i.e. the perception of character and value of information and information systems (Campbell 1996). Socio-technical approaches, therefore, are more suited to control and governance of post implementation issues, by describing and providing understandings of the relationship between technology, and organisational context and actors. Due to changing nature of interacting elements whose behaviour is unpredictable, this approach falls short for providing an all encompassing view of how to approach information systems implementation.

3.3.5   Organisational Imperative

This approach to information systems implementation is mainly attributed to the information processing model. The fundamental premise of this perspective is that strategic planning is the key to organisational effectiveness and efficiency. It argues that management has unrestricted control over choice of technology and its impact in the organisation. Organizations and the use of technology within could thus be viewed as a brain that induces fragmentation, routinisation, and binding of decision making practices that make it manageable (Morgan 1997). Organisational imperative theories in information systems are strongly influenced by strategic management theories. This influence gained momentum after Porter (1979) proposed his theory on competitive advantage. Porter's five forces industrial analysis model and related strategies have been used as basis for many research endeavours on information systems based competitive advantage (Porter and Millar 1985; Earl 1989; Galliers 1991a, 1991b).

Organisational imperative theories follow a top down approach, and generally focus on activities such as formulation of an information policy aligned with the business strategy, followed by the information architecture that is designed to cater for the overall business as well as individual business process needs. These steps thus provide a roadmap of information systems development and implementation, by taking into consideration factors such as costs involved in development and implementation of information systems, organisation's technical infrastructure, technological trends, and the risks involved in the process. In these approaches, consideration given to information systems planning outclasses information systems implementation; and implementation issues are believed to originate from the post implementation investigation of factors that hamper successful implementation (Galliers 1991a). Mintzberg (1990) criticises the top-down approach and argues that by following this approach, strategy formulation represents a controlled and mindful process that is associated exclusively with top management, and that the process of strategy formulation is isolated from its implementation. Due to this disconnect, strategy formation becomes a one way street without any feedback on its effectiveness, whereby strategy implementation processes do not inform strategy formulation processes. Davenport (1998) takes the argument further and concludes that the highly structured top down approaches do not provide an effective way of information systems implementation. The author suggests that business environment goes through continuous rate of change and these methodologies are not in keeping with the pace of this change. It must also be acknowledged that information used to formulate strategy is historic; therefore the assumptions arrived at from the analysis of this information has little relevance for future decisions. In most cases, the speed with which technology updates itself renders these strategic considerations obsolete. Consequently by the time strategy is fully implemented the primary principles adopted and assumptions made about the business are outdated, and this approach ends up strategising for the past not for the future.

These three theoretical perspectives have described the existing principles employed to implement technologies within business organisations. Each of them have their own limitations and benefits, and are further dependent on a variety of intra or extra organisational factors for their success. However, for implementation of information systems for asset management, none of these theoretical perspectives could be considered as all encompassing or all inclusive. Theoretically, a hybrid approach that draws on all these three perspectives seems most appropriate for information systems implementation for asset management. The following sections describe how information systems must be implemented to align strategic asset management considerations with technology, so as to respond to external and internal challenges.

3.4 Technology Institutionalisation and Isomorphism

There is significant theoretical support for factors influencing technology assimilation success. Diffusion of innovation (Rogers 2003) is a process in which technology is communicated through certain channels over time and within a particular social system. Rogers (2003) argues judgement of a technology adoption based on their perceptions of five attributes of it: relative advantage, compatibility, trialability, observability, and complexity. Nevertheless, technological context consists of both internal/external aspects such as equipment and processes. The organizational context embodies characteristics and resources of the organization, like managerial structure, managerial obstacles, firm's size, and degree of specialization, centralization, and formalization. The environmental context is the arena in which the firm conducts its business and concerns the size and structure of the industry, the macroeconomic context, the firm's competitors, and the regulatory environment. In sum, the way an organization sees the need for, searches for, and adopts technology is influenced by these elements.

Institutional theory focuses on the environmental factors, and offers explanation for social actions, social structure, and cultural persistence through a process by which social schemas, rules, norms, routines, and typifications (cultural beliefs and scripts) become established as authoritative guidelines for social behaviour (Powel and DiMaggio 1992).

Institutional isomorphism is a process in which organizations try to excel in their practice of social rules, ideals, and practices by fitting themselves with the environmental conditions. These institutional pressures push firms to adopt shared notions and routines. Thus, the interpretation of environment in general and technology adoption intentions in particular is affected by organisation's perception of these pressures. This process forms the base of institutional theory and neoinstitutional perspective. Coercive (constraining), normative (learning), and mimetic (cloning) are three isomorphic mechanisms which influence organizations in institutionalising technology (Powel and DiMaggio 1992). Figure 3-8 demonstrates these three institutional isomorphic mechanisms and the concepts related to each of them.

The coercive isomorphism occurs by organizational desire to conform to laws, rules, and sanctions established by institutional actors or sources, which result in gaining legitimacy and external validation that improves the organization's access to resources (Bjorck 2004). The coercive pressure is exerted on organizations by other organizations upon which they are dependent (DiMaggio and Powell 1983), for example a powerful organization can exert pressure on its partners by raising requirements such as conforming to a security standard as a condition for customer requirements. The dependant organizations will call attention to the asymmetry of power when they perceive coercive pressure and, thus, better understand the consequences of adopting or not adopting the technology. In general, the dependent organisation tends to comply with the powerful organisation's demand and be inclined to adopt and routinize technology usage into daily operation in order to maintain relationships with powerful partners (DiMaggio and Powell 1983).

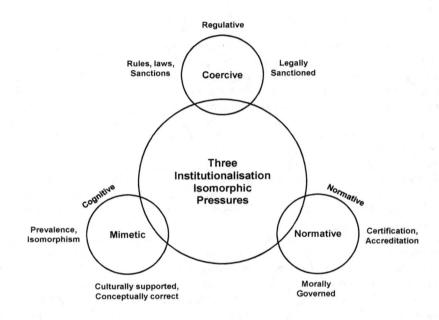

Figure 3-8: Institutional Isomorphism Mechanisms
Source: (Powel and DiMaggio 1992)

The normative mechanism concerns the moral and pragmatic aspect of legitimacy by assessing whether the organization plays its role correctly and in a desirable way. The progressive use of technology in an organization could be viewed as the result of normative influences, such as, ATM service is a standard service offering by retail banks, and banks who are not offering this service are more in the risk of damaging their legitimacy in the view of their industry and other institutions. Normative pressures evolve through organisation-supplier and organisation-customer inter-organizational channels as well as other partners, professional, and industry institutions. For instance, the frequency of technology usage among an organization's suppliers and customers may wake up decision makers' awareness of the technology and ignite organization's inclination to adopt it. Furthermore, compliance with norms with respect to environmental concerns can lead to profitability, e.g., reducing organizational cost by conforming to an environmental norm such as reduction in wastage of efforts, time, and resources (DiMaggio and Powell 1983).

Finally, the mimetic isomorphism is a cause of organizational tendency to remain similar to its peers in order to get positive evaluation from the organizational environment. This mechanism results in reducing uncertainty, improving predictability, and benchmarking other organizations that are performing at or near optimum level (Scott 2001. Organizations who are structurally equivalent and having similar economic network position, similar goals, produce, and commodities are more likely to imitate each other. Moreover, organizations mimic because they anticipate similar benefits. Therefore, when an organization starts adopting and implementing a technology, other competitors from the same industry becomes aware of it and considers adopting it. Gosain *et.al* (1997) state that the process of technology diffusion may be driven by 'need to conform and imitate' rather than just by rational decision making and technological progress. Noncompliance with each of these mechanisms comes with a risk of costly penalties, or in the worst case with the death of organization.

3.5   Alignment of Information Systems with Strategic EAM

In asset management, information systems are not just business automation tools. Among others, the most significant contributions of these systems are that they translate strategic objectives into action, and inform asset and business strategy through value added decision support. However, the fundamental building block to enable such a value profile is the quality of the alignment of strategic business objectives with the physical, social, and technical context of the organisation, such as policies, internal structures, systems, and relationships that support business execution (Scott Morton 1991). These contexts and their mutual interaction help organisational maturity, by shaping collaboration, empowerment, adaptability and learning in the organisation (Tapscott and Caston 1993). The mutual interaction of these contexts depends there critical aspects, firstly, the design of the organisation, i.e. the structure of the organisation and functions, and the reporting relationships that give shape to this structure; secondly, the business processes and related information flows; and thirdly, the skills and competencies required to execute business and operate enabling technologies, i.e. job design, training, and sourcing and management of human resources (Beaumont and Sutherland 1992).

The concept of aligning strategic business objectives with the physical, social, and technical context of the organisation illustrates that information systems implementation be aimed at binding these contexts together, so that they contribute to the strategic advantage of the business. As a result, institutionalisation of these systems contributes to the maturity of these contexts and increases organisational responsiveness to internal; as well as external challenges (Henderson and Venkatraman 1992, 1996; Earl 1993, 1996).

It has to be acknowledged that each implementation of information systems is different and it is not possible to follow particular theories (i.e. technological determinism, social technical alignment, and organisational imperatives) regarding implementation in letter and spirit. For example, information systems for asset management include operational technologies like sensors and other condition monitoring systems whose behaviour is highly predictable and require minimal human intervention. On the other hand, there are others systems like CMMS, ERP, or MIS, whose behaviour and use is determined by the social interactions of the organisational actors in the organisation. At the same time, the information demands put on information systems in some areas of engineering asset management (such as maintenance) are quite diverse and the available technologies are not mature enough to address these demands. This limits the choice of technologies and also influences their application and use. The dynamics of asset management, therefore, suggest that for effective information systems implementation there needs to be a hybrid approach that brings together social, organisational, and technical contexts of the organisation and aligns them with strategic business orientation.

There have been numerous attempts made at describing information systems alignment, however two classical approaches proposed by Earl (1996) and Henderson and Venkatraman 1992) have been the focus of practical and research endeavours. Earl (1998) while proposing his organisational fit framework (figure 3-8) suggests that alignment of technology is subjective and needs to be driven by the context rather than strategic orientation of the business. This framework attempts to propose a holistic view of information systems implementation. In doing so, the framework suggests four processes (i.e. clarification, innovation, foundation and constitution processes) that provide alignment between the four strategic domains, i.e. business strategy, information management strategy, information systems strategy, and IT strategy. Each of these domains is further divided into *components* and *imperatives*. *Components* represent the key factors that govern the domain, whereas *imperatives* illustrate the key aspects that need to be taken in account to manage the domain. This framework provides guidelines for strategic management of IT and information systems, as well as their integration.

Earl (1998) argues that the organisation must have answers to some fundamental questions to align the four domains. Although the framework does not answer these questions; however, formalises them into the strategic agenda of the organisation and indicates the processes through which these questions are raised and answered regularly. These questions could be,

a.   What information systems and IT applications the organisation should develop to improve competitiveness of their business strategies?
b.   What technological opportunities should the organisation consider to enhance efficiency and quality of its business processes?
c.   Which IT platforms should the organisation be developing and for that what plan and policies are required?
d.   What IT capabilities should the organisation develop and how may these be acquired?
e.   How should the information systems activities be organised and what is the role of information systems?
f.   How should information systems/information technologies be governed and what profile of a manager best serves these needs?

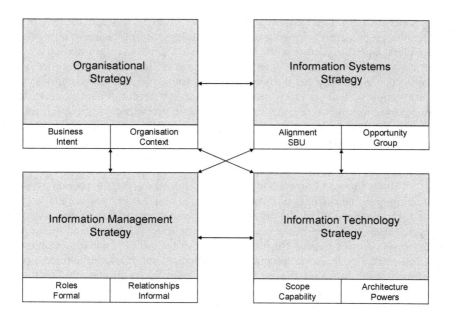

Figure 3-9: Organisation Fit Framework
Source (Earl 1998)

The framework has organisational strategy domain at the forefront and suggests its two *components* as, the organisational intent (Hamel and Prahalad 1989) interpreted through the strategic choices; and the organisational context, shaped by the organisational infrastructure and culture. The *components* and *imperatives* of the organizations' strategy need to be accounted for while formulating information systems strategy. The organisational context and business intent are subjective and, therefore, the process with which they feed into information strategy is not always clear or formalized. Earl (1998) terms understanding of these strategic considerations that influence information strategy domain as the 'clarification process', and argues that familiarity with strategic business intent and the organisational context is essential for information systems implementation and management. Information systems strategy is, thus, developed in response to this process of clarification.

The two key *components* comprising information systems strategy domain are 'alignment' and 'opportunity'. 'Alignment' is based on the 'clarification process' and calls for aligning information systems implementation with business intent, goals, and context. The 'opportunity' *component* seeks out to seize opportunities for organisational growth and maturity through creative use of technology by actively looking out for technology centric business improvement enablers and thus constituting the 'innovation process'. Information systems strategy domain influences other domains through this 'innovation process', for instance, the promise of translating or informing organisational strategy with information systems is much greater than making structural adjustments. At the same time, information systems strategy domain prompts changes to information management when reconfiguration of the functionality of these systems necessitates business process reengineering; or information systems opportunities influence technological scope of IT strategy as the innovation process necessitates acquiring new technical abilities.

The domain of IT strategy deals with two *components*, i.e. scope or types of technologies that the organisation needs to use; and the architecture that controls the technologies used by the organisation. *Imperatives* in IT strategy are capability and powers. Scope of the technological capability is determined by the skills and competencies needed for proficient use of technology, whereas architecture is influenced by the powers required to implement and manage the technological infrastructure. In doing so, the IT strategy domain constitutes the 'foundation process', which provides the base of the management and control of activities associated with building and developing IT infrastructure. The fourth domain, i.e. information management strategy, works as the bedrock of information systems strategy and comprises roles and relationships. The *components* of the information management strategy domain are the roles and relationships that need to be defined in managing IT activities, particularly the ones related to information systems' function. Roles refer to the formal associations that define responsibility and controlling of power to manage information management resources, whereas relationships define the informal relationships of the responsibility and the controlling power. The linkages that the information management strategy domain provides with information systems strategy, IT strategy, and organisation strategy domains are described as the 'constitution process'. This 'constitution process', thus, influences organisational strategy, capabilities and effectiveness of information systems strategy, and the quality of the IT related strategic decisions.

The alignment modelled in this framework provides a high level view of integrating technology with business. It describes alignments in broad terms and does not provide guidelines that could be drilled down to the operational level implementation of technology. It views alignment of information systems as a linear or a mechanistic process that follows fixed paths and interacts with 'standard' contexts. However, in reality, information systems alignment is non linear, takes time, and cannot be attained thorough an assumed set of strategies built around roles and relationships. In addition, the notion of assumptions provides a contradiction to what is proposed in the framework. Viewing alignment as a mechanical process implies deterministic stance, which affects adaptability, and also impedes creativity and novelty proposed by the innovation process associated with the information systems strategy domain.

It is important to note that values, roles, and their relationships are not just important for information management, but are equally significant for the overall alignment of technical, organisational, and social contexts. Furthermore, formal roles and relationships could be embodied in business strategy; however, human relationships that shape and influence these relationships are dynamic and thus cannot be restrained to the boundaries of a policy or a plan. The framework also stresses planning of associations between processes, rather than the relationship between technology and processes in the first instance, and then using the information thus generated to integrate business processes. This framework, thus, treats information as a passive entity in translating strategic business considerations into action, or informing business strategy so as to ensure strategic recalibration or re-orientation. It needs to be acknowledged that using information to drive alignment facilitates creation of shared meaning of the use of information systems, and helps in shaping the context within which alignment is sought. For example, information enables team work and thus aids in developing a culture favourable to the roles and relationships advocated as necessary for alignment in the framework. This framework or the theories based on this framework are, therefore, inadequate to meet the requirements of information systems implementation for asset lifecycle management.

Henderson and Venkatraman (1992) provide an alternative view of information systems alignment as illustrated in figure 3-9. The authors propose two important points, i.e. distinction of IT strategy from information systems infrastructure and processes; and distinction of strategic fit from inter domain alignment, as the key to business transformation. The model, thus, takes an intentional view of organisational transformation. It draws its value from three types of relationships; i.e. the fit that links two domains horizontally or vertically, inter domain alignment, and alignment of all the domains with strategic business considerations. It argues that business strategy consists of three key elements, i.e. the scope of the business, which relates to the services and products that the business offers; unique competencies, the attributes of the organisation that provide it with a comparative advantage over other competitors; and governance, which reflects the strategic choices, such as strategic alliances and joint ventures to support the unique competencies and the business scope.

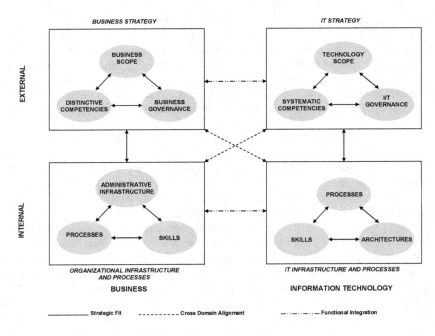

Figure 3-10: Strategic Alignment Model
Source (Henderson and Venkatraman 1992)

Henderson and Venkatraman (1992) suggest that IT strategy needs to be drawn from business strategy. In doing so, it establishes three key areas, i.e. definition of scope of IT, which illustrates the range of technical infrastructure available to the organisation; systemic competencies, which represent the distinctive IT related competencies that support existing strategy as well as contribute to the creation of new strategies; and IT governance, which are the structural choices (such as partnerships and joint ventures) to acquiring IT capabilities that contribute to systemic competencies and scope of IT in the organization.

The third domain in the model is IT infrastructure and processes, which represents the IT architecture, i.e. technological configurations and information; processes, i.e. the activities necessary to support IT operation, such as mainte-nance; and skills, i.e. the competencies required to operate and manage IT infra-structure in the organisation. Similarly, the fourth domain of organisational in-frastructure and processes represents the administrative infrastructure, including the structure, roles, and reporting relationships; processes and information flows associated with execution of key business activities; and the skills, i.e. the capa-bilities and competencies required to execute the key activities that support busi-ness strategy.

The concept of alignment demonstrated by this model is dynamic and takes into account changes in the business environment and their implications on the strategic and organizational development (Henderson and Venkatraman 1993). The clear distinction between business and IT domains advocated by this model underscores the need for functional integration, and thus calls for aligning the choices made in relation to IT and business at strategic as well as operational levels (Nelson 2001). However, the model does not account for social relation-ships that shape technology use and thus institutionalise technology. Conse-quently, the changes in IT strategy, IT infrastructure, and organisational infra-structure are in response to changes in the business environment.

This model treats IT strategy as a controlled process undertaken by top man-agement, and assumes that control of IT infrastructure, skills, and IT manage-ment processes provides the basis for technology alignment with the organisa-tional infrastructure. Furthermore, managerial action provides for integration of activities within and across domains, and in doing so the model assumes that factors like what skills are needed, how information flows between processes and systems, and what outputs will be achieved from certain control actions can be determined and hence the alignment process takes a linear path. This frame-work suffers from the same inhibitions as Earl's organisation fit framework and, therefore, is not competent enough to address the question of alignment of in-formation systems with strategic asset management so that the organisation is responsiveness to internal as well as external challenges. This framework also undermines the role of information in achieving alignment of the social, techni-cal, and organisational context with the strategic business orientation. In sum-mary, this model may be effective in analysing the impacts of information sys-tems implementation rather than facilitating asset management maturity by ena-bling alignment of strategic asset management considerations with technology implementation.

3.5.1   Information Systems for EAM Alignment Perspective

Information systems implementation and its alignment with organisational social and cultural environment, structure, infrastructure, and strategy do not follow a mechanistic pattern, and require time to take shape and deliver expected results. It is a process that is socially and technically composed in the organisation and, therefore, requires maturity of interacting actors and infrastructure to provide an appropriate level of alignment. Using available information systems theories along with the lessons learnt from the alignment theories discussed in the previous sections, this section attempts to develop an alternative approach to information systems implementation and its alignment with the technical, organisational, and social contexts of the organisation. An information systems based engineering asset management alignment framework is illustrated in figure 3-10.

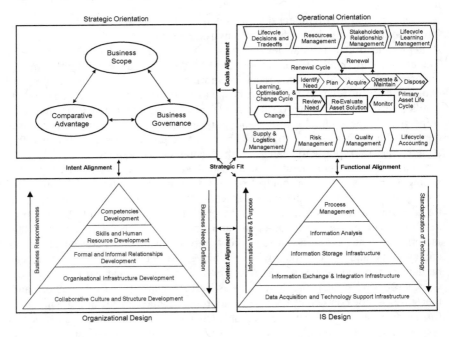

Figure 3-11: Information Systems Alignment with EAM

This framework treats alignment as a process that is technically and socially composed and embedded in the organisation; and highlights the role of information in shaping alignment. Proponents of contingency theory (Ginberg 1980; Zmud 1982; Barki *et al.* 2001; Becerra-Fernandez and Sabherwal 2001; Chin *et al.* 2003; Khazanchi 2005) suggest that performance of an entity, (for example information systems or an organisation) is contingent upon various internal and external constraints. These theorists, thus, highlight four important points, i.e. there is no one best way to manage an organisation; subsystems of an organisation need to be aligned with each other as well as the overall organisation; successful organisations are able to extend this alignment to the organisational environment; and organisational design and management must satisfy the nature and needs of the task and work groups.

Contingency theory stresses multivariate nature of organisations, and along with systems theory it assists in understanding the interrelationships within and among subsystems of an organisation (Premkumar and King 1994).The framework applies systems theory (Churchman 1994) and instead of considering the organisation's or its constituent domains' properties alone, it builds upon the relationships and understanding of the domains that collectively provide for the information systems alignment within and with the organisation. This framework embodies these relationships and applies the theory of dynamic capabilities to address the changing nature of the asset management business environment, by stressing integration, building and reconfiguration of competencies to address the changing business environment (Zahra and George 2002; Daniel and Wilson 2003).

The framework takes a resource based view and proposes four domains, i.e. strategic orientation, operational orientation, information systems design, and organisational design. Analogous to Henderson and Venkatraman's model, it argues that strategic orientation of the asset managing organisation is defined through the interaction of business scope, unique competencies, and business governance choices. Operational orientation of asset management is derived from this strategic orientation. The framework seeks to develop alignment based on goals of asset lifecycle management processes with the organisation's overall objectives. This means that asset lifecycle management processes conform to the strategic asset management orientation. The asset lifecycle management domain is strategically aligned with the organisational design domain in the sense that not only that the organisational and social context conform to asset lifecycle management objectives, but they also contribute to the responsiveness of the organisation, and in so doing help asset lifecycle management processes to adapt to changes in the internal as well as external business environment.

Information requirements of asset lifecycle processes drive information systems design. However, there is an assortment of operational technologies utilised in asset management, which this framework treats as information technologies. Thus, the alignment sought between operational orientation of asset management and information systems design is aimed at functional integration of asset lifecycle. To ensure information integration and quality, the information systems design domain takes a bottom up approach and stresses standardised data acquisition and technology support infrastructure, which facilitates information integration and communication, and consequently allows for information storage in a way that it is accessible and available throughout the organisation. This helps with information and knowledge management, and functional integration.

The analysis layer refers to both, the analysis to evaluate if the existing standard of information and information systems are meeting the process and organisational objectives (hence the strategic alignment between the information systems design domain, and strategic orientation and operational orientation domains); and the level of decision support that is required at various stages of an asset lifecycle. Quality of the asset lifecycle management processes strongly depends upon quality of information, and information quality itself is a measure of how effectively the information systems cater for the information needs of the business processes. The analysis layer, therefore, also measures the integration between information systems and business processes. However, technologies, whether information or operational, are passive entities. Their use and institutionalisation are not mechanistic processes and rely on the culture, structure, and human actors in an organisation. Therefore the framework proposes contextual alignment between information systems design and organisation design domain.

Organisational design takes time to develop and its alignment with information systems is also subjected to the same time constraints. Therefore, the organisation design domain stresses the 'development' of collaborative culture and structure as the fundamental element of organisational design. This foundation provides the building block for developing organisational infrastructure (internal structures, policies, and procedures put in place to support the strategic orientation of the business), which shapes the formal and informal relationships, and drives human resources management and skills development. Organisation design, thus, provides for the development of core competencies that aid in utilising information and operational technologies as well as executing asset management processes for the advantage of the organisation through alignment based on organisational intent (i.e. organisational vision, mission and objectives). In doing so, the social and organisational contexts contribute to strategic orientation and are themselves shaped in line with the strategic orientation. In doing so, organisational design domain improves responsiveness of the organisation, which enables the organisation to respond to changes in the business environment. At the same time, since organisational design domain is strategically aligned to the operational orientation domain, it accounts for the objectives of the overall business as well as the asset lifecycle demands and goals. It, thus, provides the context within which the information systems are employed, shaped, and institutionalised.

The context of the organisation is subject to change due to internal as well as external forces; therefore the framework suggests context based dynamic alignment between information systems design and organisational design domains. This framework treats information as the key enabler of asset management and emphasises that information systems implementation is not a managerial process or activity. In actual fact it is a social process that is continuously aimed at aligning and matching information systems capabilities with business objectives and requirements. The framework also highlights that to achieve the desired results, it is important to account for the organisational areas that influence technology implementation as well as the areas that are influenced by it. This framework, thus, uses implementation of information systems as means to translate strategic asset management objectives into operational actions by enabling asset lifecycle processes, and utilises the information generated by the execution of these processes to inform asset management strategy for strategic reorientation and recalibration. In doing so, information systems implementation becomes a generative process that helps in their institutionalisation and maturity of the technical, social, and organisational context of the organisation.

3.6   Summary

This chapter investigated the role of information systems in engineering asset management. It argued that information systems in engineering asset management have diversified roles, such as these systems, are utilised in collection, storage, and analysis of information spanning asset lifecycle processes; provide decision support capabilities through the analytic conclusions arrived at from analysis of data; and provide an integrated view of engineering asset management through processing and communication of information and thereby allow for the basis of engineering asset management functional integration. The major advantage of such an integrated view is that information systems translate strategy into action and at the same time also provide informed decision support for strategy recalibration.

Information systems implementation involves understanding of the structure of the technology, the reasons for choosing particular technology, assumptions about the context in which technology is to be used, and previous experiences with technology adoption. Information systems implementation theories can be classified into three broad categories, i.e. technology deterministic, socio-technical interactions, and organisational imperatives. However, asset management requirements from information systems illustrate that their implementation be aimed at translation of strategic objectives into action; align organisational infrastructure and resources with technology; and inform asset management strategy through value added decision support. The chapter, thus, developed a framework that aligns information systems with strategic asset management consideration. It emphasises that information systems implementation is not a managerial process or activity but a social process that is continuously aimed at aligning information systems with business requirements and objectives.

The next chapter develops the case of information systems based engineering asset management performance evaluation. It discusses information systems performance evaluation approaches and highlights the value profile of information systems based engineering asset management evaluation.

4 Information Systems Based Asset Management Evaluation

As investments in technology for asset management are increasing, evaluation of their benefits is becoming a growing concern for the asset managing organisations. This chapter examines literature to develop an understanding of performance evaluation issues in general and information systems based asset management issues in particular. The chapter thus establishes the case of information systems based engineering asset management evaluation.

The chapter commences with conceptual introduction of the nature of performance evaluation and then expands the discussion to general performance evaluation methods, information systems evaluation methods, and conceptual and operational issues involved in information systems evaluation. The insights thus gained lead into setting the foundation of information systems based engineering asset management evaluation. This chapter particularly argues that in order for the evaluation process to become a useful part of the organisation's continuous improvement agenda it is essential that focus is moved away from just fixing errors and deviations from planned course of action to challenging the strategic assumptions and considerations.

4.1   Understanding Evaluation

4.1.1   Concept of Evaluation

Since early 1990s there has been an increased research activity in development of performance measurement systems aimed at various organisational levels and areas that cover a multitude of dimensions. This increase has been fuelled by certain business development theories that promote the need for employing performance evaluation as means for performance improvement, such as constraints theory (Dettmer 1997), lean enterprise (Womack and Jones 2003), and six sigma (Pande and Holpp 2004). The activity thus generated has resulted in development of numerous models, frameworks, techniques, and methods applied in different industries with varying levels of acceptance and success. However, the discussion on the effectiveness of performance evaluation has been centred on three views. The first view suggests that businesses do well if there are integrated and well-structured performance evaluation methods in place that inform and provide management with improvement indicators (Hoque and James 2000;

Davis and Albright 2004). The second view questions the role of performance evaluation in general and individual performance evaluation methods in particular and suggests that employing console style performance evaluation methods such as balanced scorecard make little or no contribution to business performance improvement (see for example, Ittner *et al.* 2003; Neely *et al.* 2004). The third view argues that success of performance evaluation as a business management activity is highly dependent upon the approach used to implement it (Braam and Nijssen 2004).

Performance evaluation methodologies have various aims; involve a range of stakeholders; and are aimed at various stages of system, product, or organisational lifecycle. The intent of performance evaluation exercise is to provide management with a progress report on the performance of the area under investigation, so as to prompt action to address the gaps thus identified. The major objective of performance evaluation, therefore, is proactive rather than reactive management (Bitichi 1994). According to Atkinson *et al.* (1997) performance measurement serves three basic functions, i.e. to co-ordinate business management activities, to monitor the progress of the organisation, and to diagnose issues in business management and execution. Working through these functions, performance evaluation provides a roadmap for proactive organisational improvement. Meekings (1995) summarises the character of evaluation exercise by arguing that,

a. evaluation exercises should provide progressive forecasting and insights into business performance;

b. instead of being tool for management control, performance measures should be aimed at providing feedback, instituting understanding, and promoting performance improvement motivation;

c. the focus of performance evaluation needs to be based on systems thinking centred on a structured change and organisational learning, rather than setting targets aimed at fire fighting or allocation of blame; and

d. performance evaluation needs to be aligned with the organisational objectives, such that all levels of the organisation understand the process and collaborate and contribute towards continuous business improvement.

These postulates have significant relevance for information systems evaluation. Information systems evaluation is a subjective activity that is highly influenced by the context within which the information systems are employed. It involves a variety of organisational stakeholders, and a range of activities, processes, and conditions, which underscores the complexity of this exercise. As identified in chapter 1, there are certain operational and conceptual issues involved in information systems evaluation that make the realisation of an all-encompassing information systems evaluation exercise difficult.

The fundamental step towards an attempt to devise an information systems evaluation methodology is to explicitly define evaluation. Neely *et al.* (1995) suggest that performance is the measure of efficiency and effectiveness of action and performance evaluation is the process of measuring accomplishments. Here measurement deals with quantification of action and accomplishment illustrates performance. Tangen (2004) takes the argument further and contends that performance evaluation represents the set of metrics used to quantify the efficiency and effectiveness of organisational actions taken towards achieving its objectives. The efficiency and effectiveness constitute the value profile that the organisational stakeholders attach to action in an organisation. For information systems, this value profile could be financial, functional, individual, organisational, or strategic advantage (Cronk and Fitzgerald 1999). However, this value profile is different at different stages of an information systems lifecycle, for example, an ex ante or pre implementation is aimed at ascertaining cause and effect of technology institutionalisation ; whereas, ex post or post implementation evaluation may be aimed at evaluation of strategic translation as well strategic advisory role of information systems. In light of this discussion information systems evaluation could be defined as "an assessment of value profile of information systems to an organisation using appropriate measures at specific stages of information systems lifecycle towards continuous improvement of the organisation to achieve the overall business objectives". This definition embodies the objectives, measures, and process of evaluation. However, formulation of an effective methodology requires complete understanding of the how, why, what, who, where, and when of performance evaluation. Edvardsson *et al.* (1994) and Oakland (1995) elaborate these and explain that in order to develop an effective performance evaluation methodology, the organisation must resolve,

a. Why measurement is required, by being explicit about the purpose of evaluation;

b. What should be measured, by carefully working out the evaluation criteria that covers all relevant dimensions of the phenomena under evaluation;

c. How it should be measured, by devising appropriate framework for evaluation such that the evaluation provides an integrated rather than disjoint view of the performance of the phenomena under investigation;

d. When should it be measured, by deciding the timeframe and frequency of carrying out the evaluation exercise;

e. Who should measure it, by identifying the stakeholders responsible for development of the evaluation methodology, as well as those who will carry out the evaluation exercise; and

    f.   How the results should be used, by making the follow up actions clear to the internal organisation so as to motivate and involve internal staff in the process of continuous improvement.

The following sections focus on the various principles and themes of information systems evaluation in general and information systems based engineering asset management in particular. These sections develop understanding of the fundamental assumptions, questions, and beliefs that leads to a logical and rational approach to information systems based engineering asset management evaluation.

4.1.2   The What and Why of Evaluation

The fundamental question in an evaluation exercise is to ascertain the subject of evaluation as well as the dimensions of the subject that needs evaluation. For example, in information systems based engineering asset management there could be different permutations of evaluation, such as information systems enabled asset managing processes, system design, a particular system (for example ERP), quality of information, or quality of decision support that the information systems provide. Nevertheless, information systems evaluations have a broad focus, and according to Teubner (2005), may involve evaluation of,

    a.   components of information technology,
    b.   software applications,
    c.   the way IT and software are applied in information systems,
    d.   business processes supported or enabled by information systems,
    e.   information systems related development processes,
    f.   techniques, methods and tools used in development,
    g.   artefacts and models built during the development process, and
    h.   information systems related management and service processes.

The conduct of information systems evaluation largely depends upon how the organisation views technology. For example, when viewed as a process automation tool (see section 3.3.2), evaluation of information systems may be aimed at assessing the way its supports business processes, which may be measured by proper mapping of process information requirements or the speed of process execution.

Traditionally, the focus of research into information systems evaluation has been on software applications or standalone information systems. However, the trend is shifting and broadening in scope. The renewed evaluation scope includes the role, number, and effectiveness of information systems in enabling and maintaining business relationships, managing outsourcing relationships among businesses, and management of cross enterprise processes (Remenyi and Whittaker 1996; Smithson and Hirscheim 1998).

4.1.3   The How of Evaluation

Every evaluation is based on some quantitative or qualitative measurement variables. These criteria are drawn from the phenomena or from the dimensions of the phenomena that is to be evaluated. Brown *et al.* (1994) identifies six generic types of performance measures that are widely employed in evaluation exercises, i.e. customer satisfaction measures; financial measures; product/service quality measures; employee satisfaction measures; operational measures; and public responsibility measures. In information systems evaluation the generally applied generic performance measures are financial measures, such as costs of implementation; technical measures, such as response time; system usefulness attributes, such as user satisfaction; and quality of the information (DeLone and McLean 1992). Information systems, however, are social systems embedded within the organisational context and, therefore, choosing criteria that encompasses evaluation of all the information systems benefits is a difficult task. Teubner (2005) points out that these difficulties are due to a range of factors, such as,

a. *Technical Embedding.* Individual information systems components are often embedded in the overall technological infrastructure, which makes it difficult to assess the performance of these individual components. For example, while evaluating the effectiveness of a condition monitoring system, it is difficult to quantify the contribution of individual sensors.

b. *Organisational Embedding.* Information systems infrastructure is an integral part of an organisation. It influences and is influenced by a number of organisational factors, such as culture, design, and structure of the organisation. It has progressively become difficult to take the impact of information systems apart from these organisational aspects. For example, utility of an information system is not just restricted to the business process or process that it enables, but is also reflected in enabling teamwork or cross functional cooperation in the organisation. It, therefore, becomes difficult to develop evaluation criteria that truly investigate the performance of information systems.

c. *Social Construction.* The social impact of information systems is well documented, which makes it much more than just technical solutions. Change as a result of information systems affects work practices as well as the intellect and working habits of employees. However, impact of information systems on staff, social life of the organisation, and collective sense making, is intangible and difficult to measure.

d. *Social Adoption.* Information systems adoption is a social process, since their use evolves over time and depends heavily upon skills of employees and culture of the organisation. It also means that information systems may not start delivering desired results straight after their implementation. Evaluation criteria, therefore, needs to account for the time frame of information systems lifecycle within which evaluation is to be carried out.

4.1.4   The When and Where of Evaluation

Information systems evaluation can be carried out ex ante, ex post, and during operation (Walter and Spitta 2004; Doherty and King 2004; Farbey *et al.* 1999). This evaluation could be further aimed at various stages of information systems lifecycle. Depending upon the stage, information systems evaluation has different aims and objectives. In ex ante or pre implementation evaluations, performance measurement criteria are generally based on cost benefits, and perceived value that the investment may bring to the organisation. This investigation is usually carried out by functional teams, who evaluate different technological options and then chose the one that, in their view, best meets the organisational needs. The measurement criteria are basically based on the assumptions of the future use of technology as perceived by the evaluators. On the contrary, during a post implementation evaluation a report card on the information systems investment is developed. This type of evaluation is generally not conducted by the people who conduct ex ante evaluations, and therefore vulnerabilities of technology in terms of purpose and effectiveness are rarely considered. Due to technological innovation and changes in business environment, even these two factors change with time. Nevertheless, this exercise is primarily amid at financial justification of the investment made in information systems infrastructure. On the other hand, during operation evaluation is often expected to produce learnings and feedback that could be used for continuous improvement and strategic reorientation of the business. However, this form of evaluation requires long term involvement, and experience, so that the purpose, use, and fit of technology within the organisation are understood. Due to limitations of time and scope, success or failure of this form of evaluation is debateable.

4.1.5   The Who of Evaluation

This question deals with who carries out evaluation as well as who is affected by the evaluation. The subjectivity and social nature of information systems evaluation makes the choice of evaluators an important issue. Evaluation stakeholders are generally from the same organisation and, therefore, not just carryout the evaluation exercise, but are also affected by it. This makes the evaluation process dubious, due to the inherent bias of these stakeholders. On the contrary, if the external evaluators are employs, they will need substantial time to understand how the business works. At the same time, the evaluation stakeholders (i.e. the entities that are not the part of evaluation exercise but are affected by it) influence the process as well as its output. For example, managers may not be directly involved in the evaluation process, yet they have significant influence on the evaluation exercises if the evaluators are from their department. The choice of stakeholders is also affected by the time of information systems lifecycle when the evaluation is carried out. For example, for ex ante evaluation evaluators are from a diverse group such as, system developers, users, project managers, finance people, and customers; whereas ex post evaluation is generally carried out by function specific teams, information systems department, external agencies, or teams with organisation wide representation.

4.1.6 Methodologies and Tools Used

Establishing appropriate methodologies, techniques, and tools for evaluation provide the rational foundation between the evaluation measures and the effectiveness of evaluation. Due consideration to this relationship is important for the fact that information systems implementation has a direct relationship with organisational context, human behaviour, and other business management tangible and intangible structures developed around information systems. Choice of evaluation method and tools needs to account for these issues. De Toni and Tonchia (2001) state that there are five types of performance evaluation models found in literature, these are

a. Vertical or hierarchical models with an economic outlook, typified by cost and non-cost evaluations connecting ROI and productivity.

b. Balanced scorecards, where several dimensions (such as financial, learning and growth, internal business processes, and customers) are evaluated separately, which are linked together in a general way.

c. Models, which can be termed as 'frustum'', where there is a synthesis of low-level measures into more aggregated indicators, but without the scope of translating non-cost performance into financial performance.

d. Models that distinguish between internal and external performances.

e. Models that are related to the value chain.

These models have also been applied to information systems paradigm with varying success. However, information systems evaluation methodologies are qualitative as well as qualitative in nature; aim at single information systems or multitude of organisational information systems; are based on single as well as multidimensional evaluation criteria; and also consider contextual and organisational factors (Cronk and Fitzgerald 1999).

4.1.7 Using the Results of Evaluation

Using results from evaluation depends upon the feedback that performance evaluation enables. However, this feedback should not just be aimed at finding faults and errors; it needs to enable action oriented change. For example, evaluation of information systems could provide feedback in terms of the gap analysis of the desired versus actual state of their usefulness or levels of performance. The gap analysis further provides insights into the factors leading to undesired level of service. This feedback would then work as an advisory mechanism that warrants corrective actions in the areas that contribute to underperformance.

4.2   Classical Performance Evaluation

4.2.1   Aims and Objectives of Performance Evaluations

Business organisations manage their performance for a variety of reasons. The motivation to measure performance is aimed at functional areas, specific activities, and even the entire business, and cover aspects such as assessment of process efficiency and effectiveness, investments in business infrastructure, and benchmarking, to improve business management and competitiveness. A detailed review of literature (Sink 1991; Kaplan and Norton 1992, 1996, 2001, 2004; Dumond 1994; Feurer and Chaharbaghi 1995; Lebas 1995; Neely *et al.* 1995; 2002; Letza 1996; Ghalayini and Noble 1996; Atkinson *et al.* 1997; Atkinson 1998; Neely 1998; Otley 1999; Martins 2000, 2002; Dabhilakar and Bengtsson 2002; Marr *et al.* 2003; Beynon-Davies *et al.* 2004; Heemstra and Kusters 2004; Lesjak and Vehovar 2005; Tangen 2005; Standing *et al.* 2006; Wieder *et al.* 2006), yields the following rationales and motivations that organisations have to measure performance:

a. Strategic planning, which requires assessment of specific aspects of business aimed at taking stock of the existing business resources and capabilities so that management is aware of the existing potentials as well as the constraints and limitations of the organisation. This enables management to set realistic goals and objectives.

b. Change management evaluations, which aim to investigate if a newly implemented initiative, such as policy, process, procedure, or technology is working according plan.

c. Performance report card assessments, which are closely related to the change management evaluations in scope. However, these evaluations do not necessarily evaluate a new initiative. These exercises aim to measure the actual performance of a strategy, policy, or investment initiative by comparing their performance with their planned goals and objective.

d. Assurance evaluations, which evaluate effectiveness of the implementation of a policy, process, procedure, or artefact by ensuring that the assumptions made about its implementation are still applicable.

e. Compliance assessments, which are aimed at checking if the organisation is meeting its obligations, such as legal, financial, and environmental requirements of operating in a particular financial, political, or social environment.

f. Business stakeholders' evaluations, which aim at examining the effectiveness and efficiency of relationships with external business stakeholders.
g. Feedback advisories, which assess the performance of employees and departments so as to create the shared need for maintaining or upgrading the existing standard of performance.
h. Informational advisories, which provide decision support to business managers.
i. Motivational evaluations that encourage employees to exploit opportunities to meet strategic objectives of the organisation.
j. Benchmarking exercises, which aim to compare organisational performance with other organisation with a better track record in areas of concern.
k. Improvement advisories, which are learning based exercises aimed at evaluating the opportunities for growth thorough creativity and learning. These evaluations are aimed at improving responsiveness of the business, so that the organisation adapts to the changes in the business environment easily.

Evaluation methodologies, models, and frameworks based on the above types of evaluations have been employed in information systems filed with varying degree of success. These measurements have been carried out to achieve a variety of objectives. Nijland (2004) summarises these objectives as envisaged by different researches (see for example, Kumar 1990; Land *et al.* 1993; Ballantine and Stray 1998; Powell 1999). Nijland (2004, p54-55), argues that information systems evaluations aim to,

a. justify investments;
b. enable organisations to decide between competing projects (which claim the same resources);
c. enable decisions concerning expansion, improvement or the postponement of projects;
d. gain information for project planning;
e. act as a control mechanism over expenditure, benefits and the development and implementation of projects;
f. act as a learning device enabling improved appraisal and systems development to take place in the future;
g. evaluate and train personnel responsible for systems development and implementation;
h. ensure that systems continue to perform well;
i. enable decisions concerning the adaptation, modification or dismissal of information systems; and

j.   allocate (and distribute) costs and benefits to appropriate organisational departments or business units.

This description of information systems evaluation objectives reveals three important aspects. Firstly, that evaluation exercises are aimed at different stages of information systems lifecycle; secondly, that they generally provide measurements in financial terms; and thirdly, that these evaluations generally do not account for organisational and social factors. While, these objectives explain the bias towards economic benefits in information systems evaluation literature, they also highlight the important point that information systems evaluation is a continuous activity and therefore evaluation needs to be carried out at various stages of information systems lifecycle. Various authors have advocated this stance and have stressed the importance of information systems investments during its lifecycle, i.e. ex ante, ex post, and during its operation (see for example, Willcocks 1996, Ward *et al.* 1996; Frisk and Platen 2004; Gwillim 2005). The structure of evaluation remains the same, no matter what stage of information systems is being evaluated; however, they vary in the scope, dimensions, requirements, and regulation (Hirschheim and Smithson 1999).

Performance evaluation initiatives for information systems have had success limited to specific areas of business; with no one single methodology standing out as 'the methodology' that could be applied to any situation, business, area of business, or industry. As a matter of fact, there cannot be a single methodology that could serve the purpose of multifaceted performance evaluation exercise, since each situation is unique and each has its own constraints and limitations. The following sections further highlight these points and discuss some of the well-known performance evaluation models and frameworks.

4.2.2   Financial Models

Organisations employ a variety of financial performance evaluation methodologies to measure and interpret the impact and success of a project in financial terms. The common methodologies include financial models such as, payback period, return on investment (ROI), cost benefit analysis (CBA), net present value (NPV), internal rate of return (IRR), and activity based costing (ABC). However, ABC (Bruns and Kaplan 1987) has been gaining popularity among manufacturing organisations, due mainly to its distinct advantages over the traditional measures such as ROI, CBA, and NPV. ABC is a methodology that can be used to measure cost of objects (for example products) as well as performance of activities (Langfiled-Smith *et al.* 2006, p. 356).   In doing so, it calculates the indirect costs incurred and traces their relationship with business activities driving these costs, and thus ABC can identify root cause of costs more precisely than the traditional measures. Although there are researchers who argue the practical value of ABC (see for example Maskell 1991), there are others who argue that value of ABC in terms of overhead cost reduction, decision making, and continuous improvement has never been proved (Neely *et al.* 1997).

Langfiled-Smith *et al.* (2006) describe four major issues with using purely financial measures for performance evaluation. Firstly, conventional financial measures do not provide actionable learning, since they describe what has happened rather than why did that happen. Secondly, financial performance measures emphasise only on one perspective of performance, i.e. the financial perspective. Thirdly, financial performance measures provide limited guidance for future actions, since financial measures only report on the immediate financial concerns and outcomes of actions and decisions. Fourthly, financial performance may encourage actions that decrease long term shareholder and customer value, since financial measures may force managers to achieve short term financial performance at the expense of long term performance (for example managers can improve short term performance reducing expenditure on new product development, quality initiatives, information systems, human resource development (including training), and customer relationship management).

4.2.3   Balanced Scorecard
Balanced scorecard methodology has been one of the most popular and widely applied performance evaluation mechanisms. It evaluates financial as well as operational measures and describes the impact of organisational actions or decisions on satisfaction of customers, internal processes, and improvement initiatives (Kaplan and Norton 1992). Balanced scorecard is based on four key perspectives, i.e. financial, customer, internal business processes, and learning and growth (as shown in figure 4-1). However, the distinctive feature of balanced scorecard is its strategic advisory nature. Balanced scorecard methodology aims at evaluating the long terms strategic orientation of the business rather than short term actions. Balanced scorecard can be tailored to meet the needs of any business, which means any number of perspectives could be added to the scorecard's contract. However, Kaplan and Norton (1996) argue to limit the number of perspective and suggest that the original four perspectives be given broader interpretation in order to maintain the compactness of the scorecard.

Success of balanced scorecard has been reported across many industries (Hepworth 1998), yet there have also been substantial number of issues and problems identified (Kaplan and Norton 1996). The major limitation of balanced scorecard is it focuses at a high level, and therefore its findings cannot be translated into concrete operational level measures (Ghalayini *et al.* 1997). In addition to this, even though balanced scorecard is a multi-dimensional approach yet there is no provision to account for external stakeholders of the business, such as suppliers, competitors, regulators, and community (Neely *et al.* 1995; Atkinson *et al.* 1997; Brignall and Modell 2000).

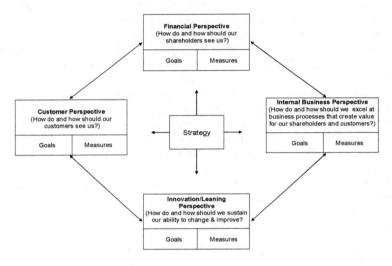

Figure 4-1: Balanced Scorecard
Source Kaplan and Norton (1996)

4.2.4   Sink and Tuttle Model

Sink and Tuttle (1989) argue that business managers need to take systems view of the organisation. Such a view describes performance evaluation through the interrelation of various systems. The authors argue that performance evaluation needs to be embedded in the strategic managerial action, such that it becomes a continuous activity and assesses performance of each managerial action. Management, therefore, needs to assess performance at the time of planning an action and then have to periodically monitor the performance of the action taken with the pre-determined goals and objectives. In so doing, according to Sink (1985), management goes through the process of,

a.   creating visions of the future;
b.   planning through taking stock of the current situation and developing action plans for improvement;
c.   creation and implementation of improvement initiatives;
d.   planning, creation, and implementation of performance evaluation systems; and
e.   enacting organisational cultural and social support infrastructure to ensure, compensate, and emphasise progress.

Sink and Tuttle (1989) propose that the performance of a system needs to be evaluated for seven categories, as shown in table 4-1. The table shows the measurement criteria for each category.

Category	Definition
Effectiveness	Checks if the system is achieving its intended goals and objectives.
Efficiency	Checks the degree to which the system utilized the correct things.
Quality	Checks if the system fulfils the requirements and expectations laid on it by its stakeholders.
Profitability	Checks if the system is giving a better return on the total cost versus total revenue ratio.
Quality of Work	Checks how do the system members respond to the cultural facets of the system?
Innovation	Checks how well does the system provide opportunities for creativity and innovation?
Productivity	Checks how good is the relationship or the transformation process that converts inputs into outputs?

Table 4-1: Sink and Tuttle's Evaluation Categories and their Description

These evaluation criteria link together to create a system that takes input, adds value to it and, provides the output in terms of enhanced level of performance. The measurement criteria proposed by this model are quite broad and can be applied to a variety of settings. However, this model is not free from issues and has some inherent limitations. Firstly, there is no suggestion on how the learnings gained from performance evaluation are to be acted upon. Secondly, the model discusses cultural affinity for performance, but does not include measures for building culture such as, human resource development, communication, and organisational structure. Thirdly, it requires explicit definition of inputs and transformation processes, which in terms of evaluation subjects such as information and information systems are not easily achievable. Fourthly, the measurement categories are subjective and context dependent, therefore, the performance criteria is bound to be different from one department to the other or one organisation to the other. The model, therefore, cannot enable all inclusive performance evaluations.

4.2.5 Performance Pyramid
Cross and Lynch (1992) view the issue of performance evaluation from a different angle and propose a sliced approach. The authors argue that since organisations consist of hierarchical structures, therefore it is necessary to adopt a pyramid approach and measure performance at each level such that all the levels contribute towards business objectives and gaols.

Performance pyramid is based on four levels and thus integrates business strategy, business units, and operations. Taking a bottom up approach the model translates the performance measures from operational level to the top or strategic level. It stresses monitoring of performance measures at the operational level, i.e. quality, delivery, cycle time, and waste, on a daily basis. Performance indicators from this level are thus communicated to the next higher levels that work as the buffer between the top and the operational levels. The performance measures employed at this level are customer satisfaction, productivity, and flexibility.

Due to the emphasis on detail, this model may suit large organisations more than small to medium sized organisations (Garengo *et al.* 2005). Nevertheless, even though on the face value the model seems to take care of a variety of external and internal factors organisational dimensions, yet it falls short on a number of accounts. For example, it does not account for the social aspects that impact performance measurement and provide information on how does the model provide for learning advisories to enable continuous improvement of the organisation (Ghalayini *et al.* 1997).

The depth of measures that the model proposes are an issue in itself, as performance evaluation is better served with the breadth of measures rather than the depth (Dickinson *et al.* 1998; McAdam 2000). The model takes a mechanistic approach and assumes objective impartiality of the evaluators as well as the passive nature of the subject of evaluation, whereas both these factors influence the outcomes of evaluation. For example, managers assessing the performance of their own department are bound to be biased. On the other hand, an evaluation subject like technology, influences many dimensions of an organisation (such as culture, design, and structure), which are difficult to measure.

4.2.6   Performance Prism

Performance prism is a relatively recent performance evaluation mechanism, and is another contribution towards multi-dimensional performance evaluation systems. Research into performance evaluation systems has generally focused on their development aspects. The effectiveness of performance evaluation mechanisms or the problems and issues associated with their application has not been given due attention (Neely *et al.* 2002). These things, however, greatly depend upon how evaluation stakeholders act and react to the performance evaluation mechanism.

Neely *et al.* (2001) recognise the significant impact that stakeholders have on the performance evaluation process and propose to include stakeholders in the evaluation mechanism. The performance prism model, therefore, includes stakeholders such as investors, customers, employees, suppliers, regulators and communities (Powell 2004). Neely *et al.* (2001) argue that businesses are following the misleading notion of deriving performance measures from business strategy, and that this approach undermines the role of strategy. It does not disregard the role of strategy in performance evaluation, instead the concept of performance prism proposes an alternative approach and stresses that a performance evaluation systems should begin with understanding what the stakeholders want. Performance prism thus proposes five factors for performance evaluation, i.e. stakeholder satisfaction, strategy, processes, capabilities, and stakeholder contribution. The model follows a linear approach of finding what the stakeholders want; how the business strategy takes care of these requirements; how the business processes contribute towards the success of the strategy; what capabilities does the business have that enable these processes; and what contributions the business stakeholders make in realising these capabilities. Neely *et al.* (2001) claim to have empirical tested performance prism in a few organisations, i.e. DHL, London Youth, and the House of Fraser with some degree of success.

The major weakness of performance prism is the attention placed on finding the right strategy for performance evaluation, rather than designing the actual performance measurement systems and the performance measures that would enable such a system to achieve the desired objectives (Tangen 2004). In addition, the model lacks in learning and innovation potential. It also falls short of explaining how the learnings from performance measurement are to be integrated with continuous improvement of the business, so that it provides for a better level or service, or sustains the existing level of stakeholder satisfaction.

4.3   The Various Types of Information Systems Evaluation

Smithson and Hirschheim (1998) indicate that information systems evaluation is carried out at five levels, i.e. macro level that includes national and international levels; sector level; company level; application level; and stakeholder level. In the first two instances, information systems evaluation research is aimed at finding the impact of information systems on businesses and society at large. These evaluations focus on determining the significance, worth, or value of information systems on the economy as well as the market dynamics. In the later three levels, the scope of information systems focuses business organisations and includes the measurement of the value profile of information systems as well as how the measurement exercise should to be carried out. Nevertheless, the major focus of information systems evaluations, both in research and industry, has been on ex ante evaluations and there has been little activity towards ex post and during operations evaluations (Frisk and Platen 2004; Gwillim 2005). In this regard, organisations seem to be more intersected in justifying information systems investments through appraisal studies rather than benefits measurement from the continuous use of information systems (Lin and Pervan 2001).

Information systems evaluations literature focus on five perspectives, i.e. financial, strategic, information systems model based, information systems attributes, and multi-dimensional. Financial approaches are aimed at financial control and enhanced resources management, and concentrate on benefits from financial feasibility of information systems rather than the ones attained from their operation and technical abilities. Strategic perspective addresses a number of dimensions that are critical for organisational integration and competitiveness, and aims to evaluate the value profile of information systems within the organisation. The information systems model perspective is aimed at enhancing the existing capabilities of information systems to provide value to the business, and focuses on improving technical aspects and functionality of information systems. Information systems attributes perspective evaluates specific attributes of information systems, such as information quality and user satisfaction, and aims to enhance the value of information systems to the organisation by improving upon these attributes. Multi-dimensional perspective involves approaches that measure the usefulness of information systems from multiple angles, such as financial, operational, and organisational. The following sections provide an analysis of the different information systems evaluation approaches.

4.3.1   Financial Perspective

These approaches are basically ex ante approaches and focus on evaluating the return (in financial terms) on information systems investment projects within the organisation. Financial approaches are based on concepts of discounted cash flow and thus measure the cash inflows as well as cash outflows associated with an information systems investment. Approaches under this category (as illustrated in Table 4-2) view information systems adoption as capital investments to facilitate business execution.

Approach	Description	References
Payback Period	Quantitative approach to calculate the time taken by investments in information systems to provide financial benefits that will cover the initial cost of investment.	Ballantine and Stray (1999); Beynon-Davies *et al.* (2004)
Cost Benefit Analysis (CBA)	Quantitative approach used to help judgements on information systems investments. CBA represents a series of different calculations with a futuristic focus that cover various aspects of information systems investments so as to enable a comparison of available information systems alternatives. It considers all costs (direct and indirect) and benefits associated with each alternative and provide investment decision support in economic terms.	Ward *et al.* (1996); Lubbe and Remenyi (1999); Heemstra and Kusters (2004)

Continued Next Page

Approach	Description	References
ROI (ROI)	Quantitative approach which compares net benefits of an information systems project against the total costs of its implementation. It focuses on the financial gains and the information systems required to generate those gains, by considering the relationship between, revenues, costs and invested capital.	Remenyi and Sherwood-Smith (1999); Stewart and Mohamed (2002); Wieder *et al.* (2006)
Net Present Value (NPV)	Quantitative approach that evaluates the difference between the present value of all cash inflows (i.e. the cost of investment) and outflows of information systems investment (i.e. economic benefits realised from the investment) using a given discount rate (minimum acceptable rate of return on investment). If the discounted cash inflow exceeds the discounted outflow, the investment is considered economically feasible.	Lee (2004); Lesjak and Vehovar (2005)
IRR (IRR)	Quantitative approach for analysing information systems investments proposals by considering the actual economic return earned by the investment over its lifecycle. Calculated through a process of trial and error, it assumes that if the IRR is greater than the required rate of return from using information systems then the investment is acceptable on financial ground.	Anandarajan and Wen (1999); Suwardy *et al.* (2003); Sharif and Irani (2006)

Table 4-2: Financial Approaches to Information Systems Evaluation

The focus of these approaches is more on the output or the contribution that information systems make in providing business value, such as cost cutting and increased productivity. Taking a deterministic view of information systems investments these approaches evaluate predetermined measures of effectiveness, and thus aim to justify investments. It is also assumed that these approaches are more suited to situations where technology is being used for process automation to cut or avoid costs (Willcocks 1992), rather than bring changes that transform the business.

Liang and Song (1994) point a significant issue with these approaches, and argue that investments decisions are made on the historical financial information of the organisation. Considering the fact that financial indicators fluctuate more rapidly than any other business phenomena, the creditability of outcomes of financial approaches to evaluate information system is debatable. Nevertheless, there are authors who recommend the use of financial approaches since these approaches are already known to the organisation (see for example Costa 1996), while there are others who argue their use after refinement such as coupling them with optimisation models (Powell 1992) or enhanced software capabilities to cover a variety of scenarios (Whiting *et al.* 1996).

4.3.2   Strategic Perspective

IT has a demonstrated potential of enhancing operational efficiency as well as changing the way a business competes (Lai *et al.* 2006). Information systems have been an integral part of the business value chain and have a critical role in enabling an organisation's competitive advantage. The effectiveness of information systems, therefore, spans from providing cost advantage to flexibility to service quality and delivery advantages (Ward *et al.* 1995). Approaches under the strategic perspective are primarily ex ante and aim to examine the cause and effect of creating value through the use of information systems (Table 4-3).

Approach	Description	References
Competitive Advantage	These approaches are aimed at examining information systems as sources of competitive advantage. They aim to evaluate strategic, tactical, operational, and financial decisions according to measures of success for the organisation.	Parker *et al.* (1988); Banker *et al.* (1993); Powell and Dent-Micallef (1997); Patel and Irani (1999)
Alignment	These approaches aim to evaluate interdisciplinary performance, by examining the relationship between information systems, operational strategies, and management practices.	Kotha and Swamidass (2000); Hemsworth *et al.* (2005); Busi and Bititci (2006)
Context Based	These approaches focus on examining the impact of information systems in particular business contexts. Thus they have a multifaceted focus and target different levels of the organisation, such as business governance, business structures, and organisational characteristics and infrastructure.	King and McAulay (1997); Shi and Bennett (2000); Zakaria *et al.* (2003); Grembergen and De Haes (2005); Klecun and Cornford (2005)
Critical Success Factors	These approaches argue that strategic success of information systems implementation depends upon certain success factors, consequently the performance of information systems is evaluated against those critical success factors. These success factors may be different from one approach to another depending upon the nature, size, and scope of the business.	Sakaguchi and Dibrell (1998); Yu (2005); Kamal (2006)

Table 4-3: Strategic Approaches to Information Systems Evaluation

These approaches argue that information systems have a major role in business design and operations, business stakeholders relationships, process and business integration, and in shaping business strategy (Bowersox and Daugherty 1995; Patterson *et al.* 2003). Therefore, information systems evaluation is not just an assessment of implementation of technology, in fact it is a critical mechanism that helps the organisation design and execute its business (Konsynski 1993), and remain competitive (Kurnia and Johnston 2000; Lee *et al.* 2003).

These approaches have a broader focus and include financial, non-financial, soft as well as hard benefits of information systems investments. These methodologies also assess the fit between information systems and different organisational strategies within specific contexts. Nevertheless, these approaches give out a simplistic notion of reality, though they are extremely difficult and complex to execute; due to the depth of performance measures that these approaches employ and their time consuming nature. However, if executed properly these approaches provide grounds for continuous improvement and thus take the organisation from simple performance evaluation to performance management (see for example Otley 1999, Schmitz and Platts 2004).

4.3.3 Model Based Perspective

Information systems evaluation research literature generally focuses on three issues, i.e. information systems utility, functionality, and usability (Willcocks 1996; Beynon-Davies *et al.* 2004). Information systems utility means, if the information delivered by the information systems provides advantage to the business; functionality stands for if the system does what it is designed for; and the usability refers to the user friendliness of the information systems. Particularly, evaluation of the functionality of the information systems is an area of concern for business as well as for the information systems developers (Beynon-Davies *et al.* 2004). However, it has been argued previously in this book that information systems are social systems and therefore their use evolves over time, as it is dependent upon the acceptance of technology by the employees, maturity of organisational and technical infrastructure, and the evolution of the organisation itself. Table 4-4 illustrates the major models and methodologies employed in evaluating the capacity of the information systems.

Approach	Description	References
Information systems implementation models	Ex ante approaches based on information systems implementation models, such as consistency model, MIT90s, and information resource management model. These approaches evaluate development and implementation of information systems against their perceived goals and objectives.	Scott Morton (1991); O'Brien and Morgan (1991); Tan (1995, 1999)

Continued Next Page

Approach	Description	References
Information systems effectiveness models	Ex post as well as ex ante approaches based on models such as information systems success model, implementation success model, and technology acceptance model, aiming to assess the impact of information systems implementation on various aspects of business.	Mahmood (1990); DeLone and McLean (1992); Venkatesh and Davis (2000); Venkatesh et al. (2003)
Maturity Models	Approaches based on evaluation of maturity models of organisational infrastructure, such as information systems infrastructure, processes, people skills and work environment. These evaluations may be aimed at seeking alignment or measuring information systems success in contributing to the business objectives.	Hinks (1998); Verweire and Berghe (2003); Saleh and Alshawi (2005)

Table 4-4: Model Based Approaches to Information Systems Evaluation

Approaches under this category aim to develop and enhance capabilities of information systems through comparison of their 'as designed' or 'as implemented' states with the 'as desired states', so that the exercise tenders a gap analysis on the originally perceived value profile of technology.

4.3.4   Information Systems Attributes Perspective

Approaches under this category examine specific attributes of information systems to assess the impact or success of information systems investments. There is no restriction on these attributes, as they range from quality of information to user satisfaction to quality of information systems to ease of use or user friendliness. These approaches (table 4-5) work on the assumption that issues occur due to specific causes, and their control results in an enhanced level of efficiency and success of the system (Guilfoyle 2000). Therefore, specific aspects of information systems are evaluated to assess their usefulness and usability, so that the evaluation uncovers the areas for improvement that provide learnings that lead to improvements in information systems design and use. For example, DeLone and McLean (1992) argue the relationship of six attributes, i.e. system quality, information quality, use, user satisfaction, individual impact, and organisational impact, on the success of an information systems. The authors argue that system quality and quality of information severally or collectively impact information systems use as well as user satisfaction. The degree of use or usability of an information system also impacts (positively or negatively) on user satisfaction and vice versa.

Approach	Description	References
Information quality	These are ex post approaches and target the quality of information contained and processed by the information systems in an organisation. Information quality is dependent on technical as well as social, human, and cultural abilities. These approaches therefore include a variety of qualitative and quantitative measures aimed at identifying the gaps in information quality that hamper the usefulness of information systems with regard to process execution, decision support, and creating value for the business.	DeLone and McLean (1992); Goodhue and Thompson (1995); Myers et al. (1997); Rafaeli and Raban (2003); DeLone and McLean (2003); Fattahi and Afshar (2006); Bose (2006)
User satisfaction/ Benefits	These approaches are largely subjective evaluations of the different users of the system, such as customers, employees, or other external stakeholders. Depending upon the nature of the evaluation it may be directed at individual, organisational, or societal level. Accordingly, the performance measures may be represent factors such as user friendliness, information availability, or benefits perceived by the stakeholders.	Ragowsky and Stern (1997); Ballantine et al. (2000) ; Wainwright et al. (2005); Gonzalez et al. (2005)
Multiple attributes	These approaches use a variety of information systems attributes to measure the usefulness or success of information systems. Selection of attributes is generally made on empirical evidence or through teamwork. These evaluations employ qualitative as well as quantitative methodologies and focus on specific outcomes that an information system enables such as productivity, and process efficiency.	Kulak et al. (2005); Hajo (2006); Ziaee et al. (2006); Standing et al. (2006);

Table 4-5: Attributes Based Approaches to Information Systems Evaluation

4.3.5   Multi-Dimensional Perspective

Information systems evaluation presents a unique set of challenges, since different types of information systems provide different types of hard and soft benefits at different stages of their lifecycle. Evaluation methodologies, thus, need to be flexible enough to accommodate for the changing nature of information systems. At the same time, the purpose of information systems and the demands that their stakeholders associate with their use also change with time. It is, therefore, important to measure the adaptability of the information systems and related infrastructure to have a measure of the responsiveness of the organisation. Approaches under the multidimensional perspective (Table 4-6) take an abstract view of the evaluation subject and assess what information systems can enable, rather than how information systems enables something. In doing so, these approaches provide multidimensional evaluate of information systems so as to provide a comprehensive performance evaluation advice. For example, Parker et al.

(1988) propose evaluation of benefits and costs of information systems through three methods, i.e. value linking, that measures benefits across functional areas; value acceleration, that assesses the value of future investment in information systems that may be necessitated by the current evaluation; and value restructuring, that measures the benefits of restructuring information systems and skills from a lower to a higher level of value.

Scope	Description	References
Analytical	Approaches that employ statistical methodology with a formal structure to arrive at a judgement. These approaches are generally aimed at analysing risks posed to or value of information systems.	Sarkis and Sundarraj (2000); Braglia *et al.* (2006)
Multi Criteria	Approaches that employ integrated multiple criteria that include financial and operational measures. Approaches under this category are based on performance measurement methodologies such as balanced scorecard, where a scoring technique is used to reveal performance of a system. However, qualitative comparisons are also used.	Bhatt (2000); Mukherji (2002); Gottschalk (2006); Atkinson (2006); Pantazi and Georgopoulos (2006)
Information Economics	Approaches that evaluate how information impacts on managerial decisions. In such approaches, economic or financial aspects of information are considered first and then the strategic criteria of performance measurement are applied. Information economics based methodologies focus address information asymmetries as well as economics of information products and the IT itself.	Mandeville (1998); Beynon-Davies *et al.* (2004); Kallinikos (2006)

Table 4-6: Multi-Dimensional Approaches to IS Evaluation

4.4   Uniqueness of Information Systems Evaluation

4.4.1   Distinct Nature of Information Systems Evaluation
Evaluation of information systems investments is unique and is different from other evaluations, due to the tangible and intangible impacts of information systems. Information systems are social systems and their interpretation is influenced by the use and meaning that organisational communities associate with them within the socio technical environment of the organisation. Evaluation, therefore, is subjected to the principles, assumptions, and concepts that the evaluators employ in carrying out the evaluation exercise. In a social setting, human interpretation is continuously evolving and thus the interpretation of information systems also reshapes due to the changes in business environment and information requirements. Evaluation, thus, represents the current meanings and

interests that individuals or communities associate with the use of information systems within the organisation.

Information systems investments are aimed at achieving various objectives, which range from enhancing the strategic and competitive advantage of the organisation (Mahmood and Mann 1993; Shaffner 1994) to increased efficiency (Busi and Bititci 2006), to sustainability of the performance (Metheny 1994; Quinn and Baily 1994). However, the measurement of the ability of information systems with regard to the contribution that they make towards these objectives cannot easily be materialised (Brynjolfsson 1994; Ray *et al.* 1994). Evaluation of information systems requires a variety of strategic organisational, economic, and social dimensions and involves external as well as internal entities and stakeholders, which makes it difficult to measure the true value profile of these systems. Furthermore, any evaluation does not serve the purpose if it does not allow or enable actionable learnings and proper follow up.

The notion of backing learnings from evaluation exercises with appropriate action also has some limitations. For example, a routine medical examination measures factors such as body mass index and blood pressure. Even if these factors fall in the ideal category, this does not mean that the person has no illness. On the other hand, if these two factors fall outside the safe limits that does not necessarily signify a disease. In order to arrive at a fitness or illness decision certain follow up actions are required, which may include further examination or employing independent measures, such as blood tests. This has particular significance for asset management; since fundamental information input into the asset health management is the condition information that originates from sensors or manual inspection.

Evaluation of this information may reveal inconsistencies, which may not necessarily be indicators of ill health. For example, in hot environment or due to continuous operation a vehicle's temperature may show higher readings; however, that does not mean that now the efficiency level of the vehicle will decline. All it signifies is that if the vehicle keeps on operating in the same context a failure may be possible. Similarly, the information systems employed for asset condition monitoring only communicate and act upon the information that is collected through sensors. The temperature or humidity sensors may report readings that have been influenced by weather. Now, these may alert to an imminent problem developing with the asset, whereas with change of weather these readings will certainly not remain the same.

Health assessment of an asset, therefore, requires a follow up action to ascertain if the problem really exists. The passive nature of information systems as decision support tools rather than decision making tools necessitates human intervention to justify follow up action. The effectiveness of the follow up action, itself, is heavily dependent upon the perspective of the decision makers, both in terms of breadth of availability of complete information and their experience and knowledge.

Business managers are faced with a paradoxical situation with regard to following up on evaluations. Cameron and Quinn (1988) argue that business managers need to balance the contrasting aspects of the paradox, by not leaning profoundly towards a particular aspect. This paradox often appears as a consequence of evaluation exercises. For example, an evaluation calling for managing consistency across different functions of a business may end up in a highly controlled and bureaucratic culture, where creativity and change may end up in culture of perpetual trialling that may impede permanence and stability. On the other hand, too much of emphasis on cross functional collaboration may result in erosion of job design, which may affect accountability and responsibility. However, this is not to argue that this paradox is entirely detrimental to the organisation. This contradiction provides the creative chaos which is important for effectiveness of the organisation, provided management has a balanced view of the organisation and the stakeholders are clear about the goals, participation, and commitment of the actions sought by management.

Important point of note is that just like the evaluation dimensions need to be comprehensive, the follow up actions also need to have a broad focus so as to eliminate the possibility of occurrence of detrimental consequences to the organisation (Cameron and Quinn 1988). Nevertheless, to control this quandary in information systems evaluation, management needs to understand the nature and impact of information systems in the organisation. Remenyi *et al.* (2000, p4) posit that there are four major areas that contribute to ineffective evaluation management of information systems, i.e. inability to, identify benefits and improvements; cover entire span of information systems in evaluation; tangible and intangible dimension of information systems; and let the information systems evolve its benefits. These areas are discussed in the following sections.

4.4.2   Inability to Identify Benefits and Improvements

One of the critical success factors of information systems investment is identification and realisation of technology benefits and the improvements that it brings to the business. The earlier this realisation can be made, the better it is for stakeholders to understand the use and institutionalisation of information systems in the organisation. This issue is, however, interlinked with the process of choosing the right technology. Often asset managing organisations, adopt a technology push strategy rather than a need pull strategy for information systems adoption (Haider and Koronios 2006), which means that technology is adopted on its reputation than on its merits of effectives in mapping the process requirements. Management, thus, takes a deterministic view of technology assumes that adoption of technology will bring perceived advantages. Right choice of technology, however, requires matching and balancing technological capabilities with organisational needs and existing infrastructure. Therefore, introduction of new technology affects organisational design as well as technological platforms.

Zmud (1988) argues that technological improvement in the organisational platform requires need pull and technology push, whereas technological improvement in the technical platform requires technology pull and need push. The success of information systems implementation lies in balancing these forces (Brown and Ross 1996), so that potential benefits are maximised through standardised technology platforms that creates a synergistic effect with the organisational infrastructure. Nevertheless, identification of potential benefits information systems investment are often elusive at implementation, as the use of information systems evolves over time. Consequently, the measurement dimensions of the benefits from information systems investments also change accordingly to cater for the existing use and effectiveness profile of the information systems.

4.4.3 Tangible and Intangible Dimensions of Information Systems

The variety of roles that information systems serve in an organisation suggests that investments in information systems provide both tangible and intangible benefits to the organisation. While, tangible benefits are relatively easy to measure, intangible measures are extremely difficult to measure. Tangible benefits represent gains in terms of reduced reliance on manpower and increased number of transaction processing.

On the other hand, intangible benefits are those that enhance intellectual capacity of the individuals as well as the organisation, quality of work environment, and culture development. A good example could be asset condition monitoring information captured by sensors. This information has high significance for asset operation, maintenance, and design. The availability and exchange of this information to these functions results in operational efficiency and also has a positive influence on the morale, motivation, and productivity of employees, for the fact that they are able to do their job easily. However, employee morale, motivation, and productivity are subjective phenomena and cannot be objectively measured. At the same time it is difficult to measure the impact of information systems in structuring, sustaining, and managing collaborative work that shapes culture of the organisation.

It is therefore difficult to ascertain the span of control of information systems. These intangible or soft factors are indispensable for the success of an organisation, however, they neither appear on the balance sheet of the business and nor are they easily measurable. Simplistic measures adopted to solely measure tangible effectiveness of information systems cannot provide exact measurements into usefulness of information systems, and renders the accuracy and credibility of evaluation mechanisms questionable.

4.5   Performance Evaluation in EAM

An extensive literature review of performance evaluation efforts in engineering enterprises from the year 2000 to 2006 (see Appendix C), reveals that a holistic evaluation of asset management processes and allied support infrastructure has never occurred. The scope of performance evaluations has generally been centred on efficiency of manufacturing strategy or production systems. Within the realm of asset management, asset maintenance and design appear to be the major concerns. However, the methodologies employed to measure performance have a function specific focus that restricts the scope of follow up actions that would contribute to improvement of asset lifecycle management. In addition, these evaluations have primarily been carried out to justify managerial actions or investment decisions. In terms of information systems, evaluations have aimed at measuring the impact of standalone systems as well as integrated systems employed at various stages of asset lifecycle and their impact on cost, quality, and throughput of business processes. Most of these evaluations have been carried out at the ex ante or ex post stages, with little research in measuring the benefits of information systems during routine day to day operations. Consequently, these organisations have failed to realise benefits from evaluations in terms of control of organisational resources, functions, and responsibilities. In these instances evaluation feedback has a limited focus and does not enforce accountability.

During ex ante evaluation engineering enterprises need to take a broad view of the technology implementation and consider the areas that may influence or be influenced by technology adoption. Researchers have generally focused on the justification of choosing particular technology and have therefore stressed on the relationship between the justification methodologies and technology adoption. In doing so, they ignore the relationship between justification of actions and the performance of technology (Small and Chen 1995). Consequently, attention is focused on the financial appraisals and technical profile of technology; rather than how the organisation might integrate technology within the existing technical infrastructure of the organisation and institutionalise it within the organisation. Although financial measures are important at this stage, but successful choice of technology is dependent on the fit of technology with the strategic and organisational environment of the organisation. It is therefore, important to consider a broad range of performance variables spanning different organisational dimensions, so as to enable an effective and comprehensive follow up based on the result/feedback of the evaluation exercise. Sun and Riis (1994) argue that in order to develop and implement a robust ex ante performance evaluation methodology, organisations need to ask three fundamental questions, which are, what specific strategic, organizational and technological issues should be considered when investing in technology, how are the strategic, organizational and technological issues interrelated to and with each other during the implementation of technology, and when should specific strategic, organizational and technological issues be addressed during the implementation process. Ex post and 'during operation' evaluations, however, have a much broader focus and are aimed at facilitating strategic action so as to facilitate continuous improvement. However, effectiveness of these actions is considerable when aims and targets of evaluation exercise are understandable and well-articulated. Evaluation, thus, becomes a learning exercise at the individual, group and organisational level.

4.5.1 Learning and Information Systems based EAM

Lifecycle learnings management is an integral part of effective asset lifecycle management (section 2.2.10), as well as of facilitating an integrated view of asset lifecycle (section 3.1.2). It should be pointed out that information systems evaluation enables a learning organisation as well as organisational learning. Both of these concepts, though distinct, are interrelated. A learning organisation is one that creates, acquires, and transfers knowledge, and transforms itself to reflect new knowledge and insights (Garvin 1993), whereas organisational learning signifies the processes though which organisations take stock of their capabilities and thus the organisations can be and are changed (Chan and Scott-Ladd 2004).

Evaluation complemented with tacit or explicit organisational action allows for the prospects of assessment at individual level, exchange of information and ideas among individuals and communities of interest, and creating consensus and agreement on the learnings from evaluation and thus ensuring commitment to the consequential follow up actions (Walsham 1993). The theory of action proposed by Argyris and Schon (1978) provides a lucid view on what makes an individual learn and what makes an individual not to learn in an organisation and the role of managers in aiding the process of learning. This theory has been the focus of much of research in organisational learning (de Geus 1996; Garvin 1993; Senge 1990). Theory of action argues that learning occurs when actors involved in the process identify and rectify disparities, variances, or faults between intentions and reality (Argyris 1996). This theory states that the intentions of actors are determined by their programs or agenda, such as the mental models argued by de Geus (1996); which set the measuring standard on which they assess the reality. Action taken to bring the situation in line with the program represents single loop learning. For example, routine inspections of an asset or periodic maintenance actions taken (such as oil change and lubrication) to keep the asset to conform to the conditions set forth by the manufacturer. Senge (1996) terms this as adaptive learning, where the actors aim to ensure conformance to an existing standard, or coping with a faulty situation with appropriate action. This approach, therefore, can be termed as action oriented conformance where specific actions and taken or behaviours are controlled to keep in line with a set of conditions.

There is another approach that targets changing the programs, agendas, mental models, rules of coping with a situation (Argyris 1996). This approach is emergent in nature and therefore provides a double loop or generative learning. For example, predictive maintenance that is aimed at detecting a failure condition in its development. In this type of maintenance the condition of an asset is continuously being evaluated against a dynamic set of condition variables and the information thus obtained is processed to continuously model the existing health status of an asset. In so doing, certain corrective actions are taken to rectify the existing problems (single loop learning), however the actions taken also become an input into the following cycles of condition monitoring (by including the results into condition monitoring software algorithms) and the new health advisories are produced by taking into consideration the corrective actions already taken. Thus the cycle of health assessment (programs, agendas, mental models, and rules) keeps on being changing, updating, and enhancing after each cycle.

Although there are certain authors who do not make a distinction between these two types of learning (see for example, Huber 1991), there are others (see for example Fiol and Lyles 1985; Senge 1990; Dodgson 1993) who strongly argue the distinction and contend that such a difference is necessary due to the aims and objectives of learning, scope of action, and change management activities. Due to the inherent behavioural and action based connotations of learning, it is often viewed as adaptive rationality. The concept of adaptive rationality is based on learning through experience, where an action generates some response from the environment and then further action is taken in response to manage the response from the environment, and hence the chain of actions and learning goes on. This perspective when extended to organisations views them as adaptive rational systems based on experiential learning. A learning organisation, therefore, facilitates learning of all its members and continuously transforms itself (Pedler *et al.* 1989).

The conception of learning has strong underpinnings in the field of psychology, and has been interpreted in many fields to generate a variety of perspectives. Ontological dimensions of organisational learning, i.e. the subject that learns, are presented in the literature at two levels, i.e. individual and collective (Curado 2006). However, organisational learning evolves from individual learning, to a shared and collective view of the whole organisation learning. The concept of networked individualism proposed by Castells (2001) provides foundation to this view, by suggesting that knowledge and information contained with and within an individual is enhanced and shaped throughout his or her life. Accordingly a learning organisation leverages and integrates individual learning in such a way that enables collective organisational learning. The process of learning, therefore, has social, technical, cultural, behavioural and organisational dimensions. This multiplicity of scope reveals the complexity of management activities to enable learning, and also illustrates the multifaceted focus of organisational learning. Wang and Ahmed (2003) posit that organisational learning focuses on five aspects, i.e. collectively of individual learning; focus on process or system; focus on culture or metaphor; knowledge management, and continuous improvement. Organisational learning, thus contributes to the culture of creativity and innovation that enables an organisation to be responsive to internal and external challenges.

Huber (1991) suggests three constructs of learning processes, i.e. knowledge acquisition, distribution, and interpretation. The author further argues that knowledge is created through information processing, and is drawn from different sources such as, the knowledge at the birth of the organisation; knowledge accumulated through experience; observing other organizations; grafting the entities that possess needed knowledge, which are not possessed by the organization; and information about the environment and performance of the organisation (p. 88). Knowledge distribution or dissemination represents the ways and means through which those who require knowledge are connected to those who possess knowledge. The concept of this connection has technical as well as social and cultural foundations. It involves information storage and knowledge management, as well as favourable and enabling organisational environment and infrastructure that facilitate knowledge sharing. Such sharing provides for the development of shared meanings and shared interpretations of organisational beliefs and objectives, which give structure to organisational memory. Organisational memory represents information acquisition, retention, and retrieval and is structured by mental and structural artefacts, i.e. individuals, culture (beliefs and mental models), transformations (process and procedures), structures (roles within an organisation), and ecology (physical setting of the organisation); and information documentation (Walsh and Ungson 1991).

Information systems for asset management have a critical role in enabling a learning organisation as well as facilitating organisational learning. Information systems based asset management evaluation therefore needs to provide insights into the effectiveness of asset lifecycle management, and also enable feedback on the relevance and fit of existing engineering asset management strategies in order to enable continuous improvement and learning. An integrated view of asset lifecycle enables management of lifecycle learnings. It allows for information integration, and thereby enables an integrated view of asset lifecycle. This view helps individuals to apply the available knowledge, and consequently learn and contribute to the intellectual capital of the organisation. For example, when asset design and reliability support information is made available to asset operation and asset maintenance, it is easy to highlight design errors and manage failures conditions, efficiency parameters, and cost efficiencies. At the same time, the information from these two functions aids in areas such as design enhancements, resource management, and assessing asset operation cost benefit.

Evaluation is thus a learning activity, which facilitates organisational learning through revealing explicit or implicit dimensions of information systems for engineering asset management. The learnings thus gained provide indicators for improvement as well as sustaining the existing form or class of asset lifecycle management. Evaluation thus works as mean of feedback on the management actions taken and their impact on the organisation (see for example, Farbey *et al.* 1993; Walsham 1993), and develop into an instrument of learning that can reduce future uncertainty of management approaches.

4.6   Summary

This chapter started with illustration of a performance evaluation methodology and an explanation of each of its characteristic. This was followed by a detailed discussion of general performance evaluation methodologies and types of information systems based performance measurements. It was argued that performance evaluation methodologies have various aims and values, involve a range of stakeholders, and are aimed at various stages of system, product, and organisational lifecycle. Information systems evaluation is significantly different from traditional performance evaluations, as it is the assessment of value profile of information systems to an organisation using appropriate measures, at a specific stage of information systems lifecycle, towards continuous improvement aimed at achieving the overall organisational objectives. Formulation of an effective information systems methodology, therefore, requires complete understanding of the how, why, what, who, where, and when of performance evaluation. Having established this understanding, the chapter then introduced information systems based engineering asset management performance evaluation and stressed that this type of evaluation is a learning activity that facilitates organisational learning through revealing explicit or implicit dimensions of information systems for engineering asset management. The learnings thus gained provide indicators for improvement as well as sustaining the existing level of service quality of asset lifecycle management strategy and plans.

The next chapter describes epistemological stance and the methodology undertaken by this research for empirical study. The research design, choice of data gathering methods, interpretation and analysis of collected data, and the limitation of the overall approach are also explained.

5 Research Methodology

This chapter presents the research methodology employed in this research. It develops the case from an overview of the research methodologies generally adopted in information systems research. Every research is built upon some fundamental philosophies, values, and viewpoints that deal with questions like, what makes up for the legitimacy of research, how to uncover the nature and sophistication of research subject or the phenomena under investigation, and what are the appropriate research techniques that could provide authentic set of evidences to bring a closure to current research. It is, however, imperative for researchers to be conscious of these inherent and veiled premises, so that they define the theoretical and conceptual assumptions governing their research in unequivocal terms (Orlikowski and Baroudi 1991).

This chapter discusses the widely adopted research concepts and paradigms in information systems research, and provides an explanation of the underlying theoretical assumption guiding this research. Previous chapters have argued the social nature of this research, thus this chapter also elaborates upon the interaction of social and technical assumptions guiding this research. Based on a review of research paradigms and their associated research methods, a qualitative interpretive approach has been chosen as the most appropriate for this research with a case study research method. This chapter describes the research process and provides justifications of choosing various assumptions and methods.

The first section discusses research paradigms relevant to information systems research and establishes the reasons for which this research has been conducted from an interpretive stance. The next section discusses the analytical lens chosen for this research, followed by the limitations of chosen research paradigm. The third section explains in detail the research strategy employed in this research by explaining the rationale and process of accomplishing each element of this research. The final section concludes the chapter and leads into the empirical study of this research.

5.1 Research Paradigms

5.1.1 Research Methods in Information Systems

Research paradigm is concerned with the methodologies, techniques, and procedures that provide empirical proof or evidence of the phenomena or questions under investigation. In a nut shell it sets the rules and conduct of inquiry (Guba and Lincoln 1989). Although, there have been numerous methods and techniques employed in information systems research, such as action research, field and laboratory experiments, surveys, case studies and simulations (Galliers 1991b), there is no single solution to the epistemological and procedural demands of the research process. As a result, the choices made by researchers often reflect the elementary observations and assumptions accepted by a community of interest that enables them to share common perceptions and undertake similar activities (Hirschheim and Klein 1989). The researcher, thus, needs to explicitly define the nature of the research paradigm that he or she employs, as it allows them to be aware of the limitations and flexibilities of the research process.

There are three paradigms commonly applied in information systems re-search, i.e. positivist or what is also referred to as conventional; constructivist or what is also referred to as interpretive; and critical paradigm. The fundamental difference between positivist and interpretive paradigms are the underlying onto-logical and epistemological assumptions. These epistemological assumptions are concerned with beliefs about what constitutes knowledge and how that knowl-edge is obtained (Myers 1997; Klein and Myers 1999). The basic ontological position of positivism is that reality is objective and can be measured. The knowledge that emerges as a result of this process is independent of the author and the measuring instruments. Thus the knowledge obtained comprises of regu-larities, causal laws, and explanations of an objective world (Ivari 1991). It also makes it clear that positivism is dominant in information systems research (Or-likowski and Baroudi 1991). However, positivistic research is often criticised for being insufficient and unsuitable in describing the impact of information systems on individual, team, and organisation levels (Lee and Liebenau 1997).

The ontological position of interpretive and critical perspectives assumes that access to reality is only gained via social constructions such as language, shared meanings, and instruments (Ivari 1991). These social constructions are not neutral and therefore have profound consequences on the object of study. These differences in ontological position (with regard to positivism) result in very different attitudes towards epistemology. Interpretive and critical perspec-tives underscore the role of human understanding, perception, and interpretation in knowledge acquisition, and thus claim that knowledge is not acquired through natural laws but from social interaction (Walsham 1993; Orlikowski and Baroudi 1991). Klein and Myers (1999) argue that information systems research is termed as interpretive when "it is assumed that our knowledge of reality is gained only through social constructions such as a language, consciousness, shared meanings, documents, tools, and other artefacts" (p. 69). In positivism, bias (which is a by product of shared meanings and common interest) is consid-ered as an obstacle to finding truth; whereas the interpretive and critical para-digms assert that knowledge and human concerns are intertwined and since re-searchers are human beings they cannot claim to be unbiased (Klein and Meyers 1999). Critical stance, when viewed from an interpretive angle, advocates the awareness of the way common perceptions and interpretations are presumed by humans. It, therefore, suggests to be mindful of the interests that these presumptions serve. Consequently, its aim is to liberate humans from intellectual as well as social control through critical evaluation of the legitimacy and reliability of opinions in knowledge creation (Lyytinen and Klein 1985).

5.1.2   Trends in Information Systems Research

One of the major characteristics of existing social research is the application of vast range of research perspectives. The variety of these perspectives is understandable if one believes that the study of social phenomena is complex and self-referential. As a result, plurality of perspectives allows better exploration of these phenomena from diverse frames of reference (Burrell and Morgan 1979). However, traditionally, in information systems research such variety of research practice is not obvious; though it is changing.

The research philosophy within the information systems research community points to the dominance of positivism (Kaplan and Duchon 1988), however recent trend is towards more socially informed perspectives. On the other hand information systems are an applied discipline rather than pure science, and, therefore, interpretive modes of enquiry are more suited to this paradigm (Galliers and Land 1987). Jarvenpaa (1988) asserts that the division between a positivist approach and an interpretivist approach to information systems research has been a theme running through information systems literature for longer than many realise. Kaplan and Duchon (1988) have highlighted that the dominant set of assumptions published in information systems research are based on a positivist approach. The authors reviewed 155 information systems empirical research articles, published over 5 years to show that much of the research conducted reflects a positivistic orientation. A further survey carried out by Orlikowski and Baroudi (1991) also indicated clear positivist dominance. Positivism accounted for 96.8% whilst interpretive studies represented only 3.2% of the studies. Trauth (1997) concludes that choosing an appropriate research method is now becoming more complex, partly because the organisational climate is becoming more challenging that increasingly questions positivist research practice. Burrell and Morgan (1979) had argued long ago that much could be gained if a plurality of research perspectives was employed to investigate information systems phenomena. Likewise, Banville and Landry (1989) emphasised that information systems research is often superficial and needs to become more eclectic, given the increasing impact of information technology on society.

The increased use of interpretive methods may also reflect a changing perspective of information systems away from technical issues to more social ones. In recent years, literature on information systems has concentrated less on the technical design of systems and more on the social nature of information systems and their development. In a time when technical artefacts are obviously of great societal importance, it is a much discussed problem as to whether technological development is following a determined and dominant path and, whether or not, in the progress of technical and social change, comparable models are perceptible. In such investigations, it is not a question of comparing two separate spheres (one social and the other technical). Instead, the approach to socio-technical analysis brings about the question: How or under which conditions can technical objects become factors of the social world?

Most IT academics and practitioners accept that information systems are primarily social rather than technical phenomena. There is a general shift away from technological to organisational issues and an emergence of more progressive accounts of systems development. Interpreting information systems in terms of social actions is becoming more popular as evidence is increasingly indicating that development and use of technology is a socially composed technical process. There are problems associated with the social and organisational aspects of technical systems that require 'understanding' not 'measurement' (Hirschheim et at. 1995; Land 1992). Perceiving information systems as social research has been slow to emerge; however, with the current change numerous socially informed approaches have been adopted now. Markus and Robey (1988) argue that each approach has its own relevant uses, for example, Mumford's (1995) socio-technical design, Gidden's (1984) structuration theory, Walsham's (1995) case study and Lyytinen's (1992) critical social theory. This lack of one clear approach might be useful in terms of epistemological flexibility but it also potentially confuses the author back into positivism. A distinction, therefore, is required to be made between methods used in social inquiry and between positivism and interpretivism. Walsham (1995) encourages researchers to adopt research strategies that focus primarily on human interpretations. There is an increase in number of interpretive researches being undertaken within information systems, which cover a range of issues (see for example, Lee 1994; Orlikowski 1993; Walsham 1993; Zuboff 1998). Denzin and Lincoln (1994b), which point out that contemporary interpretive research is experiencing unprecedented growth and expansion, though it is far from becoming the dominant paradigm in all social inquiry.

Moving beyond its traditional strongholds of anthropology, sociology and communication, interpretive research is making impressive inroads into a variety of applied fields such as computer science and information systems (Wixon and Ramey 1996), library science (Glazier and Powell 1992), health and mental health care (Morse and Fields 1995; Sherman and Reid 1994), and business and organisations (Gummesson 1991; Schwartzman 1993). However, as mentioned earlier, despite the increasing use of interpretivism, the development of these research methods has been controversial, and the debate continues on the relative merits of interpretivist versus positivist approaches to information systems research or on the possibilities of their combination (Gable 1994).

5.1.3   Interpretivism for Information Systems Based EAM

The issues concerning information systems based asset management are far too complex and multifaceted to discuss from a narrow, single dimensional perspective. These issues have technical as well as social and organisational dimensions, which not only affect the way information systems are implemented but also influence the way they are utilised for asset lifecycle management. However, as it has been discussed in the previous chapters, asset managing engineering organisations have approached information systems implementation from a deterministic point of view. Consequently, technical issues have been at the forefront of both research and industry's technical agenda. This technical school of thought represents both engineering and computer science disciplines and is concerned with the issues relating to development, implementation, and utilisation of information systems infrastructure for asset management. In so doing, they do not give due consideration to the cause and effect relationship that leads to institutionalisation of information systems in the organisation, which is itself dependent upon the shared organisational meaning of information systems through dynamic interaction of people and technology (Silverstone and Haddon 1996; Grint and Woolgar 1997; MacKenzie and Wajcman 1999).

Information systems are complex social systems (see for example Boland and Hirschheim 1987; Farbey *et al.* 1993; Walsham 1993, 1995), which cannot be detached from the context within which they are employed (Checkland 1981; Boland and Day 1989; Orlikowski and Robey 1991; Orlikowski and Baroudi 1991; Walsham 1993, 1995), they are derivatives of organisational history and human interpretation of organisational business (Deetz 1996), and also have a bearing on the technical support infrastructure. Boland and Hirschheim (1987) describe information systems research field as 'a combination of two primary fields, i.e. computer science and management, with a host of supporting disciplines, e.g. psychology, sociology, statistics, political science, economics, philosophy and mathematics. Information systems are concerned with the development of new information technologies but also with questions such as, how they can best be applied, how they should be managed, and what are their wider implications' (p. vii). Information systems discipline is, therefore, not pure science but in fact is an applied discipline (Galliers and Land 1987). However, instead of viewing information systems research from a technical or social perspective alone, it should be observed from the interaction of these two perspectives. Consequently, any attempt to understand information systems should be made through their interaction with the context within which they are employed and the role that they play in bringing about change in that context (Orlikowski 1992; Orlikowski and Baroudi 1991). As discussed in the previous section, information systems research traditionally has been dominated by positivistic perspective that treats information systems as an entity independent of social and organisational factors and their impacts. It is worth pointing out that the epistemological stance on positivism is primarily because of the research focused on assessment of technical aspects, rather than analysing and understanding the impact of technology on individuals and organisation as well as its role in business execution.

This research takes a qualitative interpretive stance that views information systems as dynamic social systems, which consist of technical, social, and organisational elements. It is through the combination of these factors that information systems are designed, utilised, controlled, altered, and managed (Angell and Smithson 1991; Cornford and Smithson 1996). Having examined the various research accomplishments of structuration theory (Walsham and Sahay 1999; Orlikowski 2000; Orlikowski and Barley 2001; Pozzebon and Pinsonneault 2005), socio-technical systems theory (Mumford 2000; Ryan *et al.* 2002; Lamb and Kling 2003; Whitworth and De Moor 2003), and actor network theory (Callon 1986; Orlikowski *et al.* 1996; Larsen *et al.* 1999; Scott and Wagner 2003), this research attempts to prevail over the one-dimensional notion of technological determinism as well as the separation of social and technical elements.

In order to address the research questions for this research, an interpretive stance provides a richer understanding of the contextually based information systems based engineering asset management issues, than the more conformist positivist approaches. Klein and Myers (1999) propose seven principles for conducting interpretive research in information systems, i.e., the fundamental principle of the hermeneutic circle; the principle of contextualization; the principle of interaction between the researchers and the subjects; the principle of abstraction and generalization; the principle of dialogical reasoning; the principle of multiple interpretations; and the principle of suspicion. By applying these principles this research examines, the context of information systems adoption in engineering asset management; how are information systems interpreted and adopted by humans in their jobs; how engineering asset management is supported by information systems and affects and is affected by organisational dynamics; and what theories could be produced through these explorations to better understand the nature of information systems based engineering asset management.

5.1.4   Limitations of Interpretivist Approach

Interpretivists argue that truth or reality is not objectively established and is actually created through a process of social interaction. The fundamental presumption in this regard dictates that when humans are placed in their own social contexts, there is a better possibility and a superior chance of adequately understanding the views and perceptions that they have of their own actions (Hussey and Hussey 1997). Interpretive studies, therefore, emphasise the value of qualitative data in revealing knowledge (Kaplan and Maxwell 1994) about the phenomena under study. Although, interpretive research is acknowledged for providing a thorough account of the research subject in its contextual depth; the end results of an interpretive study are often criticised and questioned for reliability, generalisability, validity, and legitimisation (Eisenhardt 1989; Perry 1998a).

5.1.4.1 Reliability

Reliability stands for the uniformity or dependability of results. Reliability of research findings is greater when a variety of independent methods provide the same conclusion to the issue at hand (Denzin 1970). This combination of methodologies used in studying the same phenomenon to establish the accuracy of results is termed as triangulation (Patton 1990). Nevertheless, combination of methodologies is not just restricted to data gathering or measurement, but also includes combination of design and analytical perspectives. Denzin (1978) identifies four basic types of triangulation, i.e., methods triangulation, which uses different data collection methods to study single phenomenon; theory/perspective triangulation, which uses multiple theories or perspectives to interpret the data; sources triangulation, which uses different data sources within the same method, and analyst triangulation, which uses multiple authors or analysts to review the findings of the study.

By adopting these modes, triangulation is considered to reduce systematic bias in data (Patton 1990); maximise benefits and minimise drawbacks of each method and corroborate or complement data collected across methods (Brewer and Hunter 1989); achieve complementary theoretical implications (Brennan 1992); and overcome intrinsic bias of single method, observer, and theory (Denzin 1978). Nevertheless, for enhancing the reliability of research findings Eisenhardt (1989) suggests that the researcher should begin with wider research problem, set up methodical data gathering process, and ensure greater access to generate triangulated measures. Research results from a qualitative study can be reinforced through a combination of sources such as observation, interviews, surveys, and company documents (Robie *et al.* 2006). This research makes use of interviews, observation, surveys, and company documents for reliability and accuracy of results.

5.1.4.2 Validity

Validity refers to the legality of the research findings. Qualitative research requires concrete explanatory information that leads the reader or the audience of the research to comprehension of the gist of research experience (Stake 1995). In doing so, it provides the research audience with an interpretive understanding of reality (Angen 2000). Validation in a qualitative case study based research can be achieved through the cross case comparisons, and data triangulation. Along with these, construct validity can also be achieved by ascertaining a recognisable chain of events and evidences and through getting the draft reviewed by key case participants (Remenyi *et al.* 1998). This research utilises a cross case analysis to ensure validity of results.

5.1.4.3 Generalisability

Generalisability means the degree to which the results of the research can be applied to situations other than the ones examined (Robson 2004). One of the aim of qualitative research is to provide the audience with the description of case that would enable them to replicate the process in another setting (Vaughan, 1992). Although, an individual case may not provide adequate grounds to make strong generalisations, still it sufficiently ascertains the presence of an experience that is sufficient for exploratory research (Remenyi *et al.* 1998). On the other hand, multiple case study approach provides adequate grounds for establishing generalisability. However, Stake (1980) recommends that investigation should be aimed at collecting data that has practical and purposeful implications rather than developing obscure laws, such methods of data gathering may be in conceptual harmony with the experiences of the research audience and thus provide accepted grounds for generalisation. This study is being conducted in the applied area of engineering asset management, however the findings of this research could be applied to control and management of information systems in any large sized organisation.

5.2   Research Approach, Method, and Case Selection

5.2.1   Design of Research Process

Asset management is filed of knowledge that has only recently gained attention in academic research. Information systems based asset management, however, has hardly been researched; therefore this research design is of exploratory nature. It also means that there is no validation of existing theories in this dissertation; in fact the purpose of this research is to generate new theories from empirical data. The design of this research is based on Eisenhardt's (1989) eight step iterative model (Table 5-1).

Step	Activity	Reason
Getting Started	• Definition of research question • Possibly a priori construct • Neither theory nor hypotheses	• Focuses efforts • Provides better grouping of construct measures • Retains theoretical flexibility
Selecting Cases	• Specified population • Theoretical, not random, sampling	• Constrains extraneous variation sharpens external validity • Focuses efforts on theoretically useful cases – i.e., those that replicate or extend theory by filling conceptual categories

Continued Next Page

Step	Activity	Reason
Crafting Instruments and Protocols	• Multiple data collection methods • Qualitative and quantitative data combined • Multiple investigators	• Strengthens grounding of theory by triangulation of evidence • Synergistic view of evidence • Fosters divergent perspectives and strengthens grounding
Entering the Field	• Overlap data collection and analysis, including field notes • Flexible and opportunistic data collection methods	• Speeds analysis and reveals helpful adjustments to data collection • Allows investigators to take advantage of emergent themes and unique case features
Analysing Data	• Within-case analysis • Cross-case pattern search using divergent techniques	• Gains familiarity with data and preliminary theory generation • Forces researchers to look beyond initial impressions and see evidence through multiple lenses
Shaping Hypotheses	• Iterative tabulation of evidence for each construct • Replication, not sampling, logic across cases • Search evidence for "why" behind relationships	• Sharpens construct definition, validity and measurability • Confirms, extends, and sharpens theory • Builds internal validity
Enfolding Literature	• Comparison with conflicting literature • Comparison with similar literature	• Builds internal validity, raises theoretical level, and sharpens construct definitions • Sharpens generalizability, improves construct definition, and raises theoretical level
Reaching Closure	• Theoretical saturation when possible	• Ends process when marginal improvement becomes small

Table 5-1: Process of Building Theory from Case Studies
Source: (Eisenhardt, 1989, p. 533)

5.2.2 Getting Started

This research started with the definition of the research questions that are aimed at investigation of how information systems facilitate alignment of strategic asset management considerations with overall business strategy and organisational design; what factors impact institutionalisation of information systems based engineering asset management processes and their performance evaluation; and how information systems based asset lifecycle management processes should be evaluated? Definition of research questions upfront is extremely important to specify the domain of research as well as for identification of disciplines that interact and impact the overall research domain. Researchers argue that the significance of defining research questions at the beginning is the same as in examination of hypothesis (Eisenhardt 1989). The early definition of research questions also helps in formulating the constructs that guide the research process. In case of this research, literature review in chapter 2, 3 and 4 set up the theoretical underpinnings for asset management, role of information systems in asset lifecycle management, and the evaluation of information systems based asset management. The concepts thus developed provide functional and practical foundations to carryout empirical field study. However, as with every project, things also change in research projects over time. It includes changes in research questions, constructs, and even focus (Eisenhardt 1989). Among these most common changes that occur are the changes in research focus, which occur during the iterative process of data gathering and its analysis.

5.2.3   Selecting Cases

This research is applied that was carried out in a Cooperative Research Centre (CRC) for Integrated Engineering Asset Management. The concept of a CRC brings together leading industry partners and universities to establish and execute a research agenda in particular fields of knowledge. Therefore, this research had the luxury of the availability of various organisations that could be possible subjects of study. These organisations manage a variety of assets and include semi government public sector organisations as well as private organisations. However, this research still faced some issues, such as, identifying what organisations will provide good grounds for studying the research questions. Multiple interpretive case study approach was adopted for this research.

The rationale for this approach was, firstly, there has been little research done in information systems for asset management and almost negligible for evaluation of information systems based asset management, and secondly, management of asset lifecycle is highly complex and multiple case studies help in uncovering the nature of the phenomena under study (Yin 1994). At the same time, since this research is exploratory in nature, interpretive case study method is most suitable, since interpretive case studies provide a holistic understanding of cultural systems of action, facilitate multi-perspective analysis, and include the perspective of the actors as well as that of the relevant groups of actors and the interactions between them (Yin 1994). For the case study phase of this research, organisations belonging to Australian asset managing organisations were considered. Within asset management organisations, infrastructure asset managing organisations were selected because they belong to the same area of business and, thus, provide good grounds for analysis of the research questions through analysis and cross analysis of case organisations. Initially five cases were identified for this research. However, as Patton (1990) suggests that one should take into account the objectives of research, types of research questions, and the availability of resources; it was decided that three cases would be enough to address the research issues and in the given time duration. Perry (1998b) also suggests that each case must be selected with the intention that it either produces similar results for forecasted reasons, or produces opposing results for forecasted reasons. Therefore, this research chose case organizations that manage assets for public service provision, i.e. roads infrastructure, water pumping assets, and railway tracks.

In summary, three qualitative cases were selected from Australian asset managing organisations. These exploratory case studies are to triangulate different sources for data collection, which would provide for construct validity as well as proof of results from a variety of sources.

5.2.3.1 Number of Interviews

This research adopts semi-structured interviewing technique. These interviews are conducted with key internal asset management stakeholders, i.e. such as asset planners, asset designers, asset operators, asset managers, asset maintainers, asset lifecycle support providers, and information systems managers. The rationale for selecting these stakeholders is as follows,

a. asset planners, i.e. the senior staff who are concerned with making strategic decisions about demand management, asset need assessment, and asset planning.

b. asset managers, i.e. the staff who have the overall responsibility of asset lifecycle management, innovation, and change.

c. asset designers, i.e. the staff who are concerned with design, re-design, construction, procurement, and development of asset as well as asset lifecycle support infrastructure.

d. asset operators, i.e. the staff who are concerned with asset operation, condition monitoring, and disturbance management.

e. asset maintainers, i.e. the staff who are concerned with asset maintenance and allied processes.

f. asset lifecycle support providers, i.e. the staff who are concerned with asset lifecycle support management, such as spares supply chain management, and human resource development.

g. IS/IT managers, i.e. the staff who provide IS/IT support to asset lifecycle stakeholders.

5.2.3.2 Units of Analysis

This research adopts two modes of analysis of data collected in this research, and in doing so it seeks to interpret or analyse responses of individuals, where the element or unit of analysis are the individuals or job descriptions. On the other hand, where the overall stance or philosophy of the organisations regarding asset management is compared the element of analysis is the organisation.

5.2.4   Crafting Instruments and Protocols

This research employs multiple data gathering methods. The main method is interviews with the asset management stakeholders. However, it is complemented by surveys, and direct observation. Interview method is used in interpretive as well as positivist studies Yin (1994). This method has been criticised for being based on positivist considerations, and underestimating contextual information (Boudon 1986). That is why the interviews are designed as semi-structured, interpretive, and detailed so as to enhance the subjective relevance of both the researcher and interviewees. Table 5-2 shows the various data sources employed in the field study of this research.

Instrument	Example of Sources	Types of Contribution
Interviews and Surveys	Asset Planners Asset Designers Asset Operators Asset Managers Asset Maintainers Asset Lifecycle Support Providers IS/IT Managers	Establish commonality of meaning; overview of research problem; identification of alternative aspects; disconfirming evidence; strengths, weaknesses, opportunities, and threats
Observation	Physical Setting Organisational Culture Social Interaction	▪ Evidence of IS Infrastructure ▪ Evidence of Physical symbols ▪ Cultural scenarios
Documents	Company Reports Asset Management Plans IS/IT Plans IS/IT Policies Audit Reports	▪ Credibility ▪ Interpret meanings of documents ▪ Official view of information flow ▪ Identification of themes and issues ▪ Identification of what else needs to be done ▪
Literature Review	Books Journals Conference Proceedings Industry Reports Government Reports Research Reports	▪ Identify extant theory ▪ Identify extant research ▪ Pinpoint underdeveloped issues ▪ Highlight linked issues.

Table 5-2: Multiple Data Collection Instruments
Adopted from Carter (1999)

Using multiple sources or triangulation of sources of evidence is useful in establishing convergence of results investigation (Yin 1999; Patton 1999), as well as in handling the limitations of interpretive paradigm. Therefore, the findings of a case study are more believable and credible if these are based on multiple sources of evidence. Although, interviews are helpful in understanding how humans make sense of their work, yet this method is not considered to be convincing enough to reveal how human do what they do (Barley and Kunda 2001). That is why interviews would be complemented with surveys, direct observation, and content analysis. Myers (1999) argues that in order for observation to be credible, researcher needs to spend a prolonged period of time in the field. The author further asserts that observation only works if the research field could be specifically defined, such as engineering asset management.

The static nature (water pumping assets, railway track, and road infrastructure) of the assets in the case study organisations allows for a first hand observation to review how assets are operated, maintained, and serviced. The surveys are designed for functional employees of core asset lifecycle management functions, such as asset design, maintenance, and operation. The results of these surveys would be used as add-ons to the findings from interview and observations.

5.2.4.1 Case Study Protocol

Case study protocol (Yin 1994) provides the frame of reference for the conduct of case study. It is crafted primarily for the subjects of the research. It provides a synopsis of the research so that the research stakeholders, such as interviewees, become aware of the contents, process, and conduct of research. It, therefore, includes a step by step approach including the type of questions, stakeholders to be interviewed or surveyed, and outlines the confidentially agreements and steps being taken to safeguard data and information collected through the case study. As a minimum, Yin (1994) suggests that a case study protocol must contain an overview of the research, empirical study procedures, research questions, and a brief overview of the research reports. For this research, a case study protocol was developed, which can be found at Appendix D.

5.2.4.2 Field Procedures

Being academic research, the field procedures governing this research have two parts. One part deals with the confidentiality, social responsibility, and ethics of the research; and the other part deals with the actual execution or conduct of research. Therefore, in the first instance ethics approval from the University of South Australia's research services' ethics committee was sought. This ethics approval included measures to be taken by the researcher to ensure confidentiality and secrecy of data collected, and also the undertaking to conduct the case study in a socially responsible manner. These procedures also govern the communication and correspondence with the case study stakeholders. Since most of the case study interviewees were middle managers, it was important that the visits to case study were well planned in such a way that their participation in the case study did not compromise their work.

In accordance with the University of South Australia's research services procedures an undertaking was given to case study participants about the safe and secure processing, storage, and disposal of information collected during the case study. The researcher took full responsibility of the confidentiality of research participant's responses by making every possible effort to keep the identity of the research participants confidential in reporting of this research. Research participants were also given the opportunity to request a copy of any publication that will have reference to the results of the research in which the participant participated. Names of organisations and respondents were disguised to maintain confidentiality. This was necessary because of the confidentiality arrangements of research services of University of South Australia, as well as Cooperative Research Centre for Integrated Engineering Asset Management. Therefore, organisations are identified as Case A, or Company A and respondents are identified by their job description rather than their designation.

Data collection was done through one on one interviews and surveys conducted witin asset managing engineering enterrpsies.

a.   *Case study questions*. The case study questions were carefully articulated to allow for a deeper understanding of the research questions. Several interview questions were developed, which were based upon the research questions. These questions were pilot tested within the asset management community of interest to observe possible replies. Thus a set of case study questions was formulated (Appendix D), which were used in all the cases for data collection purposes.

b.   *Survey methodology*. In addition to interviews and direct observation, this research used surveys to complement the findings from interviews and observation. These surveys were conducted in departments representing different functions of asset lifecycle, such as asset design, operation, and maintenance etc. The result of the analysis of data gathered from these surveys was used to complement the findings from the interviews conducted with the middle managers or departmental managers. Thus it was helpful for the researcher to uncover the view of technology perceived by the department as a whole, and also if there was any contradiction in the way technology and its use were viewed in the department. Xu (2003) suggests some of the advantages of surveys as, they reach a geographically dispersed sample simultaneously and at a relatively low cost; standardised questions make the responses easy to compare; they capture responses of the people that they may not be willing to reveal in a personal interview; and results are not open to different interpretations by the researcher.

5.2.4.3 Development of the Survey Instrument

Information systems based engineering asset management evaluation depends upon three dimensions, i.e. the asset lifecycle processes that the information systems enable; the elements of an information systems, such as software, hardware, information, and skills; and the value profile of asset lifecycle management, such as efficiency, effectiveness, availability, compliance, and reliability of an asset solution (see section 1.2.2). Figure 5-1 illustrates a three dimensional integrated view of information systems based engineering asset management.

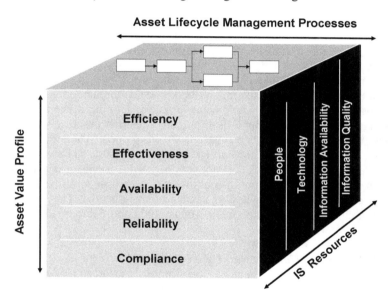

Figure 5-1: Dimensions of an Integrated IS based EAM Evaluation

The fundamental step in crafting a survey instrument to assess the effectiveness of information systems based engineering asset management is to identify the business processes that enable each stage of asset lifecycle, such as design, operation, maintenance, rehabilitation, renewal, and retirement, and thus contribute towards asset efficiency, effectiveness, availability, compliance, and reliability. In order to have complete measurement of the effectiveness of information systems in enabling these processes, different dimensions of information systems must be assessed for each of these processes, such as the capability of technology (hardware/software) in enabling the process, skills of the staff in operating technology, and the availability and quality of information generated, possessed, and processed by information systems in enabling asset lifecycle management

processes. The complete survey can be found within the case study protocol in
Appendix D.

The survey instrument measures the maturity of information systems dimensions for each process under each category of asset lifecycle management on a
scale of 1 to 5, as shown in table 5-3.

		Scale		Description
1	-	Non Existent	:	Not present
2	-	Initial/Ad hoc	:	Elementary; replete with issues and problems
3	-	Defined	:	Controlled state; however there are issues
4	-	Managed	:	Managed state; however does not provide added value
5	-	Optimized	:	Economical, value added, and continuously improving

Table 5-3: Scale of Information Systems Performance

5.2.5   Entering the Field

The major flexibility of Eisenhardt's (1989) model is the iterations between data
gathering and data analysis, which helps in theory building. Generally researchers struggle to detach data analysis from data gathering process, mainly due to
the fact that data gathering is a time consuming process. Therefore, the model
suggests that a researcher could start analysing data as soon as he or she has sufficient data to begin analysis. This not only highlights room for improvement in
the case study instruments, but also allows the researcher to refine the findings
through contacting the interviewees again to clarify details. The fieldwork for
this research was conducted between August 2005 and February 2007, and followed the following strategy.

5.2.5.1 Contact and Conduct with Case Participants

As mentioned earlier, the case study organisations are part of an Australian government sponsored Cooperative Research Centre for Integrated Engineering Asset Management or the CRC IEAM. The researcher had to overcome some bureaucratic issues to contact the case organisations. Firstly, the researcher had to
contact the education and training manager of CRC IEAM with a choice of organisations that were deemed desirable for the empirical studies. The education
and training manger then arranged for the preliminary meeting or a telephonic/email contact between the researcher and the contact person in the case
study organisation. The researcher then had to present a brief outline of the research to the contact person in the case organisation and a decision about
whether the organisation was willing to participate in the research was communicated to the researcher within a week's time. Once agreement was received a
formal case study protocol and letter of support from the researcher's supervisors
were forwarded to the contact person in the case study organisation. It is worth
mentioning that generally the contact person was the asset manager or business
development manager in the case study organisation. The documentation sent to

the case study organisation included the number of interviews and surveys sought, as well as the job description with whom the interviews were sought. Upon agreement on the interviews, a formal contact was established with the persons representing each job description. Agreement was also sought about the conduct of survey within the departments of the interviewees.

It is obvious that the issues relating to information systems investments in asset lifecycle management are multifaceted and require a broad and flexible perspective for comprehensive examination. It includes investigation of technical as well as an assortment of others dimensions such as organisational, social, and cultural. In order to address the research questions, around two dozen middle managers representing various roles such as asset designers, maintenance engineers, network access manager, business development manager, Operations and Maintenance manager, manager projects, manager assets management, project officer assets, finance manager, and IT managers associated with asset lifecycle management were interviewed. Interviewees were chosen on the basis of their responsibilities as they are between senior mangers (who make decisions) and operational employees (who act on the decisions made by senior mangers). They are the actual implementers of information systems and, therefore, are well placed to provide insights into policy setting and decision making of the senior management and the issues and challenges posed to these policies and decisions at the operational level.

The researcher had extensive telephonic conversations with the contact person in each case organisation, during which the researcher explained important issues like the time, duration, place, and conduct of interviews. Generally, the place of interview was the interviewee's office and the time duration ranged between one to one and a half hours. The interviews were semi structured and the interviewee had the option of withdrawing from the interview or research at any time. The researcher explained the confidentiality and secrecy of the information collected to each interviewee and survey participant before they committed to participate in the research. In addition an information sheet was also provided to each research participant. During the interviews a relax approach was adopted, whereby the researcher started with general questions to 'break the ice'. Permission to record the interview was, however, sought formally. Interviewees were encouraged to express their opinion freely. However, it was the interviewee's responses that drove the sequence of questions than the order identified in the case study protocol (Carson *et al.* 2001). This helped in engaging with the interviewees in a free and candid expression of their views on technology as well as its use in the organisation. The interviews were followed up by email and telephone for further clarifications, where it was deemed necessary.

5.2.5.2 The Case Study Analysis Procedures

This research employs qualitative interpretive stance in the three exploratory case studies. It is not hypothetic, and is open ended that seeks to provide assistance to management. It puts a methodical structure on the already existing knowledge from literature as well as practical world. Qualitative approach allows for investigation of nature of elements and uncovers new elements through their interrelationships. Within qualitative research, interpretive perspective allows for understanding a phenomenon through the connotations and meanings that humans associate with it. Interpretive perspective allows for "producing an understanding of the context of the information system, and the process whereby the information systems influence and are influenced by the context" (Walsham 1993, p. 4-5). Its focus is on complexity of asset management, its associated processes within the business, and the way these processes realise the asset management as well as overall business strategy. In asset management context information systems are not only facilitating the realisation of asset lifecycle management processes, but at the same time are also informing the business strategy through information processing and decision support. This makes asset management a highly complex and intricate phenomenon.

Qualitative research in the form of case studies is considered when the issue under investigation is highly complex, and unquantifiable (Yin 1994). Yin (1994) further adds that, it is especially helpful in uncovering the 'why' and 'how' questions associated with a phenomenon. Walsham (1993, p. 15) states that "the validity of an extrapolation from an individual case or cases depends on the plausibility and cogency of the logical reasoning used in describing the results from the case studies and in drawing conclusions from them". Case study approach, thus, provides better explanations of the entity under examination as it allows for rich descriptions (Miles and Huberman 1994), which would otherwise be lost in experimental and other quantitative designs. Nevertheless, qualitative methods utilise large amount of data and the major challenge lies in making sense and drawing inferences from it (Hussey and Hussey 1997). Yin (994) describes data analysis as a process of examining, categorising and presenting data, such that it reveals patterns that are helpful in resolving the issues relating to the research phenomena under investigation.

As mentioned before, data captured in this research has been gathered by three different methods, i.e. interview responses, surveys, and direct observations formed by the researcher after examining the workflow and company documents. These data were organised using three different ways. Interview responses were transcribed and were tested for accuracy through a couple of run-throughs by comparing the recording with the transcriptions. Survey data was compiled in simple excel spreadsheets and the observations were recorded and categorised in a separate file with each observation recorded under corresponding category, such as asset design, operation, and maintenance etc. For analysis of interview data NUDIST (Non-Numerical, Unstructured Data Indexing, Searching and Theorizing) software was used, where data was coded and categorised according to the categories identified from literature. However, these categories were not exhaustive and other categories, identified during analysis, were incorporated. The main categories of codes used for this research are, organisational structure, technological infrastructure, asset design, operation, maintenance, operational efficiency, competitiveness, and information systems evaluation. NUDIST helped the researcher in creating indexes of these categories in a cascading tree format. With this type of a relationship it is easy to establish relationships between different categories of data and their connections in complementing and supplementing sub categories or main categories. In this way information relating to a category or a sub category can be extracted quickly and can be examined in its entirety. It also helps in highlighting important issues through ongoing analysis of information, and its facility of memo helps in generating notes on the run during analysis. These notes help a lot in generating theoretical assertions.

5.2.6   Data Analysis

Data collected through semi structured interviews, surveys, and observation was analysed at two levels. Firstly, information systems utilization in each stage of asset lifecycle was analysed and then the same was analysed within the overall context of the information systems based asset lifecycle management in the organization. Secondly, information systems utilization in the individual cases was compared against other organizations to obtain consensus on different aspects of technology use. The analysis phase started with analysis of each case individually to reveal how the organisation under investigation implemented information systems for asset lifecycle management, what influence these systems have on enabling each asset lifecycle stage's business processes, and the impact of these systems on the value chain of asset lifecycle management as well as on the organisation.

Once information systems utilization for each case was identified, cross case analysis was carried out to further verify and validate the findings. Individual case analysis or what is also referred to as within-case analysis started with the analysis of each organization. This included information about competitive position of the business within industry; the structure of industry; and internal structures, such as asset management strategy, information systems strategy, and human resources strategy. It helped in understanding the various decisions and postures that the organization took in adopting and utilization information technologies in general and information systems in particular. This background knowledge made it clear to understand the broad organisational rationales of information systems utilisation for asset lifecycle as well as the role of information in enabling asset lifecycle processes. The themes and patterns revealed through the interviews, surveys, and content analysis helped develop a deeper understanding of how information systems manage each stage of asset lifecycle, what role they play in terms of enabling an integrated information enabled asset lifecycle view, and whether the case organisation evaluated its information systems based asset management processes that aimed at continuous improvement. The within case analysis for each individual case included the analysis of data from the interviews. Following Patton (1990) quotes from case study participants have been used to highlight the insight gained to address the issues under observation. These qualitative assertions were further complemented with the survey data to enhance the validity and further highlight particular points. At the same time, wherever applicable, instances recorded through personal observations were also used to confirm or reject certain points. Conclusions were thus drawn to interpret the patterns involved in use and institutionalization of technology in case organisations. Quite often in qualitative studies the researcher faces the issues of divergence, where it appears to be a contradiction in particular points. These issues were addressed through cross case analysis by studying those factors in different settings. This process not only enhanced reliability of findings but also helped in validation of certain points.

Nevertheless, cross case analysis revealed common factors and patterns from all the cases. Those common patterns allowed for a richer understanding of the asset management issues faced by the case organizations, and at the same time helped in understanding the ways in which these organizations have arranged there information systems infrastructure to respond to the internal as well as external business demands.

5.2.7   Shaping Hypotheses

This step in the model represents the process where the researcher aims to find alignment between theory and the collected data. The extent of this alignment defines the convergence of construct and thereby promotes its validity. In this research original theoretical foundations were set forth from literature review in chapter 2, 3, and 4. These theoretical foundations were continuously finetuned and reformed through the data collected during the case studies. In so doing, qualitative data proved to be extremely helpful in explaining the 'how' and 'why' of the evaluation of information systems based asset management issues. As a matter of fact, several new arguments came to forefront through this iteration process, which will be presented in chapter 6 and 7.

5.2.8   Enfolding Literature

Eisenhardt (1989) argues that a vital characteristic of theory development is the evaluation of evolving ideas, postulates, or theories with existing literature. The author further contends that such an alignment between theory and literature enhances generalisability and legitimacy of theory. However, as has been discussed in this research, literature in information systems for asset management and evaluation of information systems based asset management is very limited. Therefore, the literary underpinnings for comparison extend over a wide range of disciplines and make it practically impossible to seek consensus across the board. However, theories developed through this research and the findings reached could be applied to any public sector large organisation engaged in developing and/or continuous improvement of its technological base and its alignment with business strategy.

5.2.9   Reaching Closure

At this stage the researcher is faced with question of judgement about theoretical saturation. This saturation could be defined in terms of iterations between data collected and theory, as well as the collection of data. According to Eisenhardt (1989), a researcher may be motivated to stop due to time and budgetary limitations. However in this research, it was the time constraints as well as the practical ability to carry out the case study within the available financial and operational resources that dictated the closure of the research process. At the same time, the other factors necessitating closure was the level of clarity, consistency, and depth of findings, which are discussed in chapter 7.

5.3   Summary

This chapter described the scope of research and the approach taken to undertake this research. The chapter started with an in-depth discussion of popular research paradigms in information systems research. This was followed by a discussion of the appropriate choice of approach for this research. It argued that the issues concerning information systems based asset management and its evaluation are quite diverse and unique, and thus qualitative interpretive approach with a case study methodology is most appropriate for this research. It particularly argued that socially constructed realities consist of social meanings that cannot be defined objectively or precisely. Hermeneutic and dialectic processes are, therefore, required through language exchange in their own unique epistemological and practical perspectives. However, there are certain limitations of an interpretive approach, and therefore in order to enhance the validity and generalisability of this research triangulation of data sources was chosen. The main method of data collection is the semi-structured interviews; however, it is complemented with observation and survey methods.

Nevertheless, the research process is based on Eisenhardt's model of theory development with a case study method. Following this, each of the 8 steps of the model and the way these have been applied in the research has been discussed in detail. In summary, this research framework has the potential to provide rich insights and interpretations for development of information systems based asset management evaluation theory. The next chapter provides an illustration of this and explains the three case studies conducted by utilising this research framework.

6   Case Studies - Information Systems Based Asset Management

Chapter 5 described the research methodology and data collection methods for this research. This chapter reports on the three case studies conducted within Australian asset managing organisations. Each case is examined as a whole in this chapter to obtain an understanding of the opinions and perspectives of the respondents from each organisation. In addition to discussing information systems utilised for asset management, each case study provides an overview of the industry that the organisation operates in, the asset management initiatives undertaken by the industry at large, asset management initiatives of the organisation, and technological infrastructure utilised to enable asset management in the organisation.

The case studies follow a narrative approach to describe and evaluate the maturity of information systems based asset lifecycle management in the organisation. Direct quotations from the case study interviews are used to complement the overall point of view being presented. Apart from this, surveys conducted in each case organisation to reveal the perception of employees towards information systems in each core asset management function have been used to highlight the maturity of technological infrastructure.

This chapter has three sections. The first section provides the case of an Australian water infrastructure asset managing organisation; the second case describes the information systems based asset management of civil infrastructure being undertaken as collaborative BOOT (build, operate, own and transfer) projects by civil engineering organisations; and the third section presents the case of rail track asset managing organisation.

6.1   Case Organisations and Analysis and Display of Data

All case study interviews together with the additional documents obtained from the case study organisations as described in Chapter 5 were transcribed and entered into NUDIST. An intensive content analysis of those documents and interview transcripts was conducted. Direct quotations from the case study interview transcripts (Patton 1990) regarding various themes of information systems based asset management have been used in this chapter to illustrate the factors or sub-factors, which could assist in building explanations (Miles and Huberman 1994) of the various aspects of the research questions at hand. Quotations from case study interviewees represented their own opinions, perceptions, and experiences regarding particular factors or situations. They also provide the respondents' true feelings and beliefs on certain issues. Therefore, these quotes have the potential to assist readers in obtaining insights into the respondents' understanding of the phenomena. However, quotes are identified by the case name and the respondent's position title.

As discussed in Chapter 5, as per the research ethics requirements, confidentiality agreements signed with the participating organisations and individual interviewees state that these entities must not be identified by their real names and/or actual position titles. Every feasible and plausible effort has been made to conceal the identifies, thus the cases are referred to as Case A, Case B, and Case C, and interviewees are referred to by their job description rather than their actual designation, for example design manager, operations manager. Number of employees in the organisation and annual revenue has also been rounded.

6.2　Water Infrastructure Asset Management – Case A

The Company A is a government owned corporation and operates in a competitive marketplace on an equal commercial footing with private sector providers. It has an operating budget of over 100 million Australian dollars per annum, and operates one of the largest state's bulk water supply and distribution infrastructure located throughout regional areas of the state. It supplies about 40% of the water used commercially in the state via 27 water supply schemes. Water supply customers number over 5,500 and include irrigators, water boards, local governments, power stations and mining, industrial and manufacturing companies. The water infrastructure assets that the company owns had replacement value of $3.8 billion (as at June 30, 2006) and consists of,

    a.　26 major dams (collective capacity of 6,314,000 mega litres),
    b.　82 weirs and barrages with collective capacity of 356,500 mega litres,
    c.　72 major pump stations,
    d.　more than 2,500 km of pipelines and open channels, and
    e.　more than 730 km of drainage works.

In addition to managing and operating these assets, the company also provides a variety of engineering and infrastructure management services to water industry in Australia as well as overseas. The company's structure consists of four Strategic Business Units, i.e. water supply services, engineering services, operations and maintenance, and corporate (figure 6-1). Asset lifecycle management spans all the four SBUs, however is more focused within operations and maintenance.

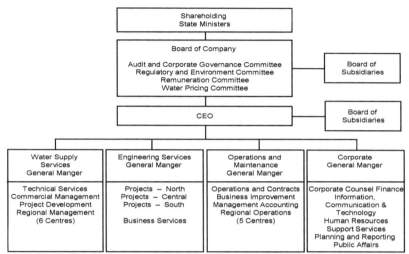

Figure 6-1: Company A's Organisational Structure

6.2.1   Australian Water Industry

Australian water industry is primarily developed around a linear model of collecting, storing, treating, distributing, and discharging water (Barton group 2005). The industry represents a mixed picture and consists of public and private water services providers. It includes federal as well as state departments that manage water resource and infrastructure through a range of policies and legislations in areas such as pricing, service levels, and environment. However, the focus of these administrative bodies is on performance improvement, water demand management, and waste management.

According to a study conducted by the Australian National University based Barton group (2005), the annual turnover of water industry is well over 8 Billion Australian dollars; though the accumulative economic impact extends well beyond this figure as it supports industry as well as agriculture sector that accounts for more than half of Australia's GDP and constitutes over 80% of its exports. The water consumption according to specific sectors is illustrated in the figure 6-2 below.

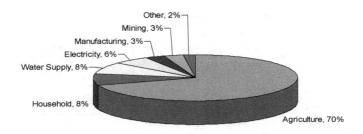

Figure 6-2: Water Usage in Australia
Source (Barton group 2005)

The scarcity of water resources and the continuing drought in some parts of Australia has placed renewed demands on water infrastructure asset management. These demands, on one hand require effective supply and demand management of water, and on the other hand require operators to sustain and manage their water harvesting assets with reduced levels of performance enforced by climatic changes. In order to counter such challenges state governments as well as the Council of Australian Governments (COAG) have engaged in reforms aimed at better planning and allocation of water. For example, the water act of 2000 embodies a variety of such requirements, and also calls for operators to prepare strategic asset management plans as key tool for enhancing asset management approaches.

The concept of Asset management in water industry has been strongly endorsed by the governments at federal as well as state levels, and water utilities have actively engaged in developing asset management regimes since early 1990s. The increased interest of engineering enterprises towards asset management is largely motivated by incentives of improvements from financial management, recovering full cost of service, and using cost benefit analysis on a regular basis to evaluate performance of their assets with the profits that they enable (GAO 2004). However, water infrastructure assets in Australia are aging fast and are exposed to challenges of various types, starting from underutilisation of assets due to drought to inability to proactively managing these assets due to vast area of their deployment. Table 6-1 summarises the results from the Australian infrastructure report card on a scale of A, B, C, D, and F, with A representing very good and F representing inadequate. It clearly shows that the water infrastructure asserts are far from being in optimum condition.

Asset Type	Australia 2005	NSW 2003	VIC 2005	QLD 2004	WA 2005	SA 2005	TAS 2005	NT 2005
Wastewater	C+	C-	B	C+	B-	C	D+	C
Potable Water	B-	C	B	B	B-	C+	D+	B-
Stormwater	C-	D	C-	C	C+	D	C-	C+
Irrigation	C-		D	C+	C+			B
Overall	C	C-	B-	C+	C+	C-	D	C+

Table 6-1: Condition of Water Infrastructure Assets in Australia
Source (Australian Infrastructure Report Card 2005)

6.2.2   Asset Management at Company A

Asset base at Company A includes dams, water stations, pumps, and water pipelines. The company, in conformance with the state government's legislation, has developed a strategic asset management plan. This plan has been in place for a number of years and is updated and revised in line with the outcomes of periodic interval review. The latest audit of the strategic asset management plan was carried out by independent external auditors in the year 2005-06. This audit covered areas such as content, reporting, standard of customer service, and technical ability. However, the technical ability audit was only restricted to the technological ability of the asset infrastructure, and did not include the evaluation of enabling or support infrastructure such as information systems. Nevertheless, Company A's asset management scope includes,

a. development and continuous upgrading of asset management strategy that defines actions such as asset identification and valuation, strategic planning, asset condition assessment, risk evaluation, asset renewal identification, maintenance optimisation, writing of operations and maintenance manuals, implementing operations and maintenance strategies and monitoring the performance of these strategies;

b. strategic asset management plan that is reviewed on a 3 yearly basis and includes a 30 year renewals program;

c. 3 yearly strategic audit plan, which is approved by the Audit and Corporate Governance Committee;

d. environmental management through AS/NZS ISO14001;

e. production of annual corporate plan with a five year outlook;

f. risk management processes based on AS/NZS 4360:1999.

Company A manages the investment and renewal process of infrastructure through the company's strategic asset management process, which is also a legislative requirement under the Water Act of 2000. The strategic asset management plan embodies planning and implementation structure for asset management activities, such as condition assessments, operations and maintenance manuals, renewals, performance/service standards and emergency action plans. Using the same plan Company A completed a 15.4 million Australian dollars program of renewals work in the year 2002-03. During the same time period the company also spent 5.5 million Australian dollars on investigation and development of new commercially viable asset infrastructure. During the year 2005-06 the company spent 9.8 million Australian dollars on asset refurbishment and augmentation plan to ensure that all water schemes were in good operating condition. In addition, 1.4 million Australian dollars were spent on preventive maintenance work bringing the total spending since 2000 to 1.2 million dollars. Company A generates the revenue for carrying out infrastructure renewals and maintenance through water sales. Asset managing organisations are increasingly investing in IT to enable asset management regimes. However, Australian Infrastructure Report Card (2001) reports that asset managing organisations generally struggle to maintain timeliness, consistency, and complements of data. Although the report card illustrates some good examples of quality information and statistics by Water Services Association of Australia; yet it has generally not been possible to obtain quality data that would enable detailed analysis of asset valuation, replacement values and asset management systems within sectors, such as stormwater and irrigation (Australian Infrastructure Report Card 2001).

6.2.3   Information Technologies at Company A

Owing to the legislations, changes in technology, and demands of aging asset infrastructure, in 1990 company A took initial steps towards introducing information systems for asset management. As a result an asset register and management system (ARMS) system was implanted. However, this system was nothing more than simple record keeping of asset inventory. Apart from limited functionality the system was not integrated with any other organisational information system, hence there was no way of finding out costs incurred during asset lifecycle. In addition, there were issues with data quality, duplication of data, and the lack of data standardisation, therefore maintenance history varied greatly from reality and lacked creditability. In 1995 the company expanded ARMS to include extra functionality such as, accrual accounting, asset identification, asset valuation, bill of quantities, and direct and indirect costs. In 2000, Company A adopted SAP R/3 as its core asset management system, and the idea is to integrate a plethora of asset management specific operational and informational technologies with it. However, the initial focus was on implementing the plant maintenance module to provide improved data quality, work management and costing, and management decision support tools. SAP was implemented due to the regulatory pressures rather than in response to the process needs of company A. In other words, the technology was introduced to the organisation without taking in account its compatibility with the existing technological architecture and infrastructure. It, therefore, became difficult for the organisation to adapt to technology and fully institutionalised it within the organisation. Consequently, implementation of SAP has been far from satisfactory. Other critical factors that contributed to this issue have been 'data migration' and 'data integration'. ARMS was not a completely functional system and relied in a number of ad-hoc databases and spreadsheets to meet the information requirements. In addition, ARMS and SAP conform to different information models; therefore, migration of data from ARMS to SAP was not possible. The company aspires to integrate these systems as well as their electronic data management systems and make it available on the company's intranet (see figure 6-3) in the next 5 years.

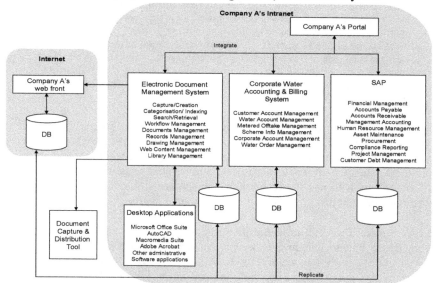

Figure 6-3: Intended Asset Management IS Architecture of Company A

The chosen SAP configuration is still not fully operational and the company is using specific modules of the system. In another major technology implementation initiative, in 2002 an Oracle based information management system was introduced to incorporate customer relationship management, and customer and water account management.

At the moment, customer billing system is available online; tough it has been sublet to a third party. In addition to these technologies, the company also utilises a variation of CMMS (with limited functionality) and SCADA systems. Company A does not have standardised technology adoption policy throughout the organisation, and it does not conform to a common information model. As a result, different functions have implemented their own customised spreadsheets and databases to aid their day to day operations. Although these applications aid in the execution of work within the department, however they are not integrated with other information systems and are of little value to other departments who could use this information for better asset lifecycle management. Company A has a strong cost focus and little commitment from senior management in terms of adoption of new technology, which is reflected by Company A's Manager Business IT systems who suggested that,

> Last year (while the company was considering implementation business intelligence infrastructure) vendor walked into the door and asked what you want. Management did not know what are business intelligence systems, what are their capabilities, and what kind of reports can they generate......... I am not a qualified engineer so I cannot quality assure if the system is capable of providing what the engineers want ....(for implementation of technology) we consider the total cost of ownership. The actual implementation may be cheaper but when we consider costs incurred on learning new technology, and costs relating to adoption of technology you have to make a decision on the total cost of ownership and not just the initial implementation cost.

In summary, Company A treats business enabling technology implementation and related areas as a secondary activity and makes no attempt to align it with asset management objectives. The ad-hoc nature of technological solutions and their management result in a variety and number of issues that do not allow the organisation to use the technology for a more strategic, planning, or management role than just process automation to some degree.

6.2.4   Asset Design and Reliability Support

Asset demand management at Company A is governed by the system of prioritising customers as well as water. However, due to draught condition in Australia, Company A's assets have been underutilised in the last decade. Most of the assets owned by the Company A were designed and developed before 1950, therefore almost all information regarding their design, except for some relating to their refurbishment is in a hard copy format. There is no exchange of design information with other asset lifecycle management functions. The lack of availability of digital information poses a number of issues, such as inability to provide decision support due to lack of integrated information that would help in holistic planning and management. However, there has been little effort made by the organisation to inculcate information driven work environment or to develop an information culture. For example, the design manager noted,

> We have particular needs in design and most of our information is driven from top down. To be honest with you our experience with technology has not been that good. Software implementation is usually very difficult to achieve and infect we've seen quite a few of them come and go without providing benefit to

us. The intricacies of integration with other applications just ended up proving to be too difficult. SAP has been with us for a long time, but we are not seeing the benefits though I am sure what SAP is capable of providing us.

Chief Engineer (Case A)

In the formal process of asset design/redesign chief engineers visit different regions and talk to designers and operations to discuss the design and operational requirements. Deign/redesign process of an upgrade or refurbishment is carried out through consultation with designers and is fully reviewed by the technical service engineers in that region. The company thus ensures that they have consensus from all the parties involved. In doing so, there is heavy reliance on the tacit knowledge held by staff. The organisation is not technologically mature enough to preserve this knowledge. At the same time, this knowledge or information is not shared with other functions within the organisation. A possible cause of this is the apparent lack of information ownership and accountability. People recording or capturing information do not understand the value of information and do so to merely keep a record of the task accomplished by them. They do not consider that the information needs to be used and reused by various other parties involved in the asset lifecycle management. This situation is reflected in the response of a design manager, who argues that

We use AutoCAD but that just gives us an electronic version of a drawing. Ideally we would like to have access to information to analyse how good our design is. The information available to us has been input by the people who really don't have sufficient technical background to understand the key things that need to be there. We've got long way to go I think before our systems are going to be sufficiently up to date and have sufficient useful information that our guys could pretty quickly get a hand on. Once we cannot get our hands on this information we have no choice but to rely on the knowledge held with our filed staff.

Design Engineer 1 (Case A)

The design exercise has a strict focus on the design of the asset and there is not enough consideration given to supportability analysis of asset design. Although it can be attributed to the fairly stable nature of water infrastructure assets, however major cause of the inability to carry out a comprehensive set of supportability analysis is non availability and lack of access to requisite information. Company A, being a semi government organisation, retains the hierarchical silo approach that resists information exchange and collaboration, which is evident from the following response of a design engineer,

Getting the historical information in order to get the design right at the first place is very difficult. In order to get this information costly onsite information extraction is required, that is by talking to the local people. We would like to see accurate maintenance information in the field, like what is happening, why it happened, what are the failures etc. On top of this, there is no interaction between electrical and mechanical engineers. They just log information in their own systems for someone else to look at and make decisions. If you want to know performance of a particular system, your best bet is to go and speak to a field engineer.

Design Engineer 2 (Case A)

This argument is further supplemented by another design engineer, which further amplifies the manifestation of a silo approach to asset lifecycle management.

We are not into risk assessments in a big way. Obviously all of systems are subject to corporate risk assessment. From design point of view we look at what condition we need to implement to manage risks posed to an asset, and that's where all the issues are that we want to monitor. Once the asset is in operation its for the operations and maintenance people to do risk assessments. They never ask us for information on any previous assessments and even if they do we cannot provide it to them easily since we perform our assessments manually.

Design Engineer 3 (Case A)

A survey was conducted in design function to ascertain if there were any divergence of views between management and operational employees of Company A (figure 6-4).

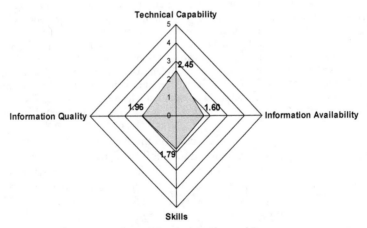

Figure 6-4: IS for Asset Design at Case A

The survey revealed the general lack of trust in capability of technology as it scored an overall mean of 2.45. The mean score of 1.60 for availability of information to aid in design/redesign activities indicates that existing information systems provide little value to execution of design processes. It highlights the issues of lack of information integration, as well as the inability to capture information at all. Since, there is no emphasis on information ownership and accountability in the organisation, it is not surprising to note that overall staff members in the design function do not feel competent to operate technology and are not comfortable with the user friendliness of technology, hence the overall mean score of 1.79. Information quality is directly affected by all these factors and although it shows a slightly elevated score as compared with skills and availability, yet the score of 1.96 is far from satisfactory.

6.2.5   Asset Operation

The nature of water infrastructure asset operation is quite different from other assets. Water is sourced from specific supply points and thus cannot be pumped from anywhere, which means that the infrastructure is static and the environmental and operational constraints on the asset are relatively easy to predict. This also means that the water asset infrastructure has to operate at a certain level and the usual principles of load apportionment do not work in this situation. There are no established systems for capacity management, only ad hoc solutions.

Company A's assets base consists of a variety of asset types and are spread anywhere between 30 km to 100 km or even longer in each direction. Asset monitoring therefore is not only costly but is also time consuming. Although, Company A makes use of GIS (geographical information system) and SCADA system to monitor asset operation, yet asset operation and condition assessment is largely manual. The information captured through SCADA systems is only used for alarms generation and failure reporting, it is not used or aggregated with other information for analysis such as failure root cause. While describing this, the operations manager of Company A suggests that,

> Condition assessment is manual exercise at the moment, since we are struggling to integrate different systems with our major asset management system (SAP). When we are required to do condition assessment, our guys will go and do that and in the process if they identify something that in their opinion presents an undesirable outcome they will flag that.
>
> Operations Manager (Company A)

Although, asset operation is the least automated area in Company A, however the company is making progress towards establishing an operations specific module within its core asset management technology of SAP R/3. According to operations manager,

> We are developing a fairly significant module within SAP, which will allow us to capture risk and condition information on each of our assets. And also to identify refurbishment and maintenance work that needs to be carried out on each of those asset. So we can actually develop a 30 - 35 year schedule to rank and prioritize vulnerabilities or risks posed to assets.
>
> Operations Manager (Company A)

They survey conducted with the staff associated with asset operation (figure 6-5) reveals the lack of automation of operations function and reveals that information systems usage in terms of capability, information quality and availability, and skills is performing below par.

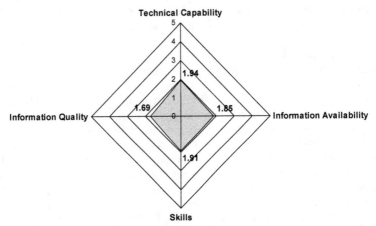

Figure 6-5: IS for Asset Operation Support at Case A

A major factor behind these figures is non standardised asset condition assessment. The multiplicity of the asset condition assessment methods (both at device and process level) hampers information integration and aggregation, which is an important building block of an integrated view of health assessment and management. At the same time reactive rather than proactive approach to condition assessment has led to the culture of 'waiting for something to happen before action could be taken'. Consequently, the function lacks a quality culture with regard to technology adoption and utilisation.

6.2.6   Asset Maintenance

Company A generally carries out a periodic preventive maintenance on its asset base, which ranges from the ones built in 1920s to date. Therefore, the maintenance demands of these assets are quite divergent. Nevertheless, maintenance information is generally processed either manually, or through an array of custom made spreadsheets developed by regional offices. Maintenance scheduling, however, is done centrally at the company headquarters by using SAP PM (plant management) module. Maintenance plans are developed with a 12 months' time horizon and include a list of tasks to be performed. At the start of each month, monthly work requests are released, and a budget allocated for carrying out these maintenance activities. Therefore, there is little provision of emergency repairs.

For example if the failure at the station requires replacement of small component or a minor treatment; it is attended to by the maintenance crew at the station. In case of major failure, maintenance requires approval from various levels as well as commissioning of expertise and resources and therefore takes time.

Company A differentiates between maintenance and asset ownership, i.e. work is executed by maintenance crew, whereas asset ownership is the mandate of another function. Consequently, there are multiple versions of the same information within the organization. Furthermore, these versions have their own bias and standard of quality. Although the organisation is aware of the potential of quality information, yet there is little emphasis made on recording and capturing correct and complete information. For example, a maintenance engineer summarised the situation as,

> Maintenance crew is not technically qualified or capable to operate an IT system. They consider it as an add-on to maintenance work, something that just has to be recorded. At the end of the day they will not be judged on what information they entered. Their performance is evaluated on the quality of their maintenance work.
>
> Maintenance Engineer (Company A)

Maintenance information, however, is crucial for asset lifecycle management, as it provides the basis for lifecycle cost benefit analysis, remnant lifecycle calculations, as well as for asset refurbishment, upgrade, and overhauls. However, like other functions, maintenance information is not exchanged with other lifecycle functions. In addition, the main focus is on capturing maintenance execution information with little provision for integrating this information with financial information. Consequently, there is no way of calculating the cost of failure as well as real costs incurred on maintaining the asset. Company A's finance manager noted that,

> There is fixed maintenance cost which is the routine day to day maintenance, and then there is what we call renewals program or refurbishment program. This is what you would call the irregular lumpy parts of your maintenance over the life cycle of the asset. We have always separated out refurbishment or renewals program from maintenance program. Most organizations will clump it together because you have to clump it together to get any kind of resource planning, but yes, we are not using the financial indicators as input into reinvestment or investment in assets.
>
> Finance Manager (Company A)

Case A relies heavily upon the tacit knowledge and expertise of maintenance crew to execute maintenance, with little help available from the information systems infrastructure except for maintenance/repair scheduling and planning activities. Although the company has invested in a state of the art SCADA system, yet the information captured by this system is not used for any diagnostic or prognostic purposes. On the other hand the company has been capturing its maintenance history from various paper based as well as customised spreadsheets into SAP. The company has centralised data entry operations for maintenance, whereby maintenance crew cannot enter information directly into the system but instead have to provide their paper based record of maintenance information to the regional office where the maintenance coordinator enters the information into SAP. This step has been taken to address the quality of information, however, has resulted in huge delays and the information available in SAP in never current. Consequently, the company cannot process the same to have an up to date set of learnings about asset condition assessment or operation profile that would allow for better management of asset health.

> Our current structure prevents us from (asset operation) profiling or managing lifecycle learnings. You also have to take into consideration that our asset infrastructure is more than 75 years old and there are aspects of assets about which we have no information available. Apart from this it also requires relevant information from other areas. We cannot profile an asset's operation through maintenance information alone.
>
> Maintenance Manager (Company A)

The figure 6-6 presents the consolidated view of the results of the survey conducted among the maintenance staff to assess the effectiveness of information systems utilised for maintenance management. The survey demonstrates that the maintenance crew believe that they do not have access to relevant information and whatever is available lacks credibility.

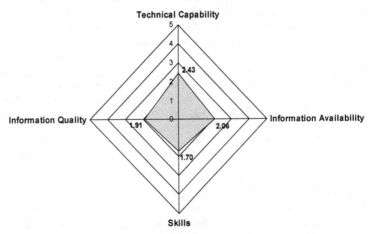

Figure 6-6: IS for Asset Maintenance and Support at Case A

The slightly raised score of information availability as compared with other factors signifies that maintainers do believe that the organisation captures the required information. However, the apparent lack of use of this information in maintenance execution has led to a low score for information quality. As a result, maintainers have little confidence in technology. In the absence of a formal training plan, there is no motivation for them to learn about the existing technologies, which is reflected through the overall mean score of 1.70.

6.2.7 Operational Efficiency

Information systems in Company A are primarily being used for recording information or what could be best described as recording what the organisation has done. This information is seldom used for more high profile purposes, such as organisational integration, planning, and executive decision support. The prevailing silo approach has thus affected departmental efficiency as well as functional integration. Management at Company A takes a deterministic view of technology and aims more at the perceived benefits from technology than the cause and effect relationships that enable these benefits. User training has traditionally been a weak area at Company A. Little training is provided and even that is aimed at training managers and supervisors rather than the staff who use the system on daily basis. The idea is that the supervisory staff (with the help of IT department) train functional staff. In these circumstances it is obvious that staff do not feel comfortable with using major technological platforms such as SAP, and business units in Company A are more inclined towards using internally developed ad-hoc spreadsheets and databases. Apart from this, there is little information exchange internally as well as with business partners, there is substantial information and knowledge drain.

Senior management is not technology savvy and therefore does extensively use IT for asset lifecycle decision support. It also explains why the organisation has been slow to adapt or even consider business intelligence or other executive level decision support tools. Even otherwise, organisational information lacks quality and there is no way of managing the important asset lifecycle learnings. Company A depends a lot on the tacit knowledge (particularly for design and maintenance) and at present more than 65% of the staff at Company A are within 10 years of retirement age, which means substantial amount of intellectual capital loss. Furthermore, the inability to provide rationale for adoption of technology to employees has resulted in functional staff considering data entry as an 'additional workload' to their job description. A senior manager from Company A attributes this to the culture of the company and summarises that,

> It has a lot to do with culture. Our culture is wrestling with fundamental issues. Some would argue that we are in an asset based industry and not an intellectual property based industry or anything like that. Certainly true to say that there is a difference of opinion in the organisation as to how asset portfolio should be

managed through IT. Perhaps people are not trained or skilled for the organisation to change. IT implementation needs to be addressed a little more strategically. As an organisation we have to try to convert people from break down heroes into more strategic thinkers.

Group General Manager (Company A)

These issues are further evident from the survey results in figure 6-7. Although the survey reveals a relatively high approval for technical capability of the information systems, however lack of training, and unavailability and quality of information has not helped them in achieving significant benefits.

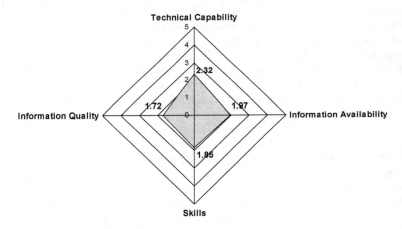

Figure 6-7: IS for Asset Management Operational Efficiency at Case A

The survey also reflects the propensity of Company A's management to use information systems as mere record keeping rather than as an enabler of tasks such as, easing and facilitating organisational workflow, enabling cross functional teams, and integration within the organisation and with business partners. This is the same reason that employees in Company A acknowledge the potential of technology but are not big advocates of its capabilities.

6.2.8   Competitiveness

Information systems in Company A are not making any significant contribution towards enabling a competitive asset management regime. At no stage asset information is integrated with financial information, which could provide assessments of asset operation and lifecycle cost benefits or efficiency statements. However, the fundamental impediment in technology not being able to competitiveness of asset management strategy is the lack of fit or alignment between technology and asset management goals and objectives. Company A is not using technology to translate asset management strategy into action and as a result the available technology cannot contribute to strategic asset management by informing management about the performance gaps through information analyses. The use of technology for asset management in Company A is evolving and is maturing along the gamut of standalone technologies. Similar response is evident from the survey (figure 6-8), which was conducted among middle to senior management of Company A.

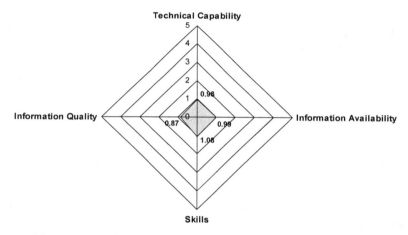

Figure 6-8: IS for Competitiveness of Asset Management at Case A

It shows lack of trust and capability of technology in improving the competitiveness of asset management initiatives in the organisation. The overall mean score is less than 1 for all the categories and it points to the fact that Company A has not fully realised the potential of information systems for asset lifecycle management. At the same time, the organisation lacks requisite planning and management of information systems implementation. Technology has been pushed into the organisation rather than being driven by the process information/users/stakeholders' needs. As a result, not only that the use of information systems has not evolved beyond simple repositories for record keeping, but the organisation has also failed to institutionalise technology in its culture.

6.2.9   Evaluation Initiatives
Company A's IT and information systems plan is reviewed every 18-24 months. However this is an informal review and is done by departmental/IT managers and is more of a qualitative observation exercise rather than a detailed efficiency assessment of information systems. It must be pointed out that this review is or-

ganisation wide and not solely aimed at information systems for asset management. Nevertheless, even in this review, water supply services group has the greatest influence, which renders the exercise biased. This is primarily due to the fact that water supply services group is the asset owner and other groups are service providers. A more detailed review of asset management was carried out in the year 2004 by an external consultant. This review was primarily aimed at assessment of asset management processes, and in the process it also reviewed information systems utilized in enabling these processes. The rationale for this review was to reorganise the company in terms of resources, accountabilities, and responsibilities.

An important driver within that reorganisation was about skills development, knowledge retention, and ensuring the availability of right environment for information exchange and knowledge sharing. The net result was a gap analysis against industry best practice. Information systems evaluation, therefore, was a on small part of the review and not the major focus of the review. This qualitative review was moderated by a steering committee of Company A, which ranked and prioritized the recommendations or the follow ups of the review. However, it took Company A more than a year to actually agree on the prioritization of the work to be done.

Company A is a traditional semi-government organisation and retain much of the character of a public sector organisation. Performance evaluation in such circumstances is a political exercise and lacks proper commitment for action oriented follow up. It is amply reflected in the observation of a company A's project manager, when he notes that,

> We have taken a step towards taking stock of our asset management related information systems resources; however there are issues that we must overcome. Firstly, the lack of ownership of data that does not provide any motivation for staff to capture and process right and complete information; secondly, investment in mobile data acquisition solutions such that data is entered as close to its origin as possible; and thirdly, effective change management with emphasis on creating a learning environment and proper training.
>
> Project Manager (Company A)

6.3   Road Infrastructure Asset Management – Case B

Case B presents the study of two organisations engaged in two Build Own Operate and Transfer (BOOT) civil infrastructure asset design and development projects. First of the two projects is a road tunnel project being undertaken in one of the largest state capitals of Australia. This project involves design and construction of twin 3.6 km tunnels linking a road to the freeway and upgrading of a further 3.4km of the Freeway. This project has been commissioned by the state government, which appointed company ABC (a consortium of companies) to finance, design, construct, maintain and operate the tunnel for 33 years, after which it will be returned to the Roads and Traffic Authority (RTA). The construction cost is the project is 1.2 billion Australian dollars. Company ABC is a contracted consortium which also includes the above mentioned joint venture of two leading civil engineering companies. The project involves 250 senior staff including engineering design, construction, urban design, environmental experts, and community relations staff based at the project office as well as at 8 site offices for construction staff.

The joint venture of the two companies that are the part of the consortium ABC's project, are also collaborating in another project (in a separate consortium) in another state capital of Australia. This project is also a BOOT project to construct roads, which will be toll taxed. This project is worth 2.5 billion Australian dollars and involves construction and operation of a toll road from the state capital to a country town. It involves the construction of 45 km of predominantly three-lane capacity freeway-standard road, 16 interchanges, 90 bridges, twin three lane 1.5km long tunnels and 35km of extensively landscaped bicycle and pedestrian paths. The project team, comprising of up to 2500 staff employed over the life of the project, includes engineering design, construction, urban design, environmental experts and community relations staff.

6.3.1   Australian Construction Industry

In the Australian infrastructure report card of 2001, ratings of Australian roads ranged between C and D. In the Australian Infrastructure Report Card (2005) these ratings have been reported as, C+ for national roads, C for state roads, and C- for local roads. This means that there has been marginal improvement in the condition of road infrastructure since 2001, and it also implies that the overall condition of roads can best be described as only 'adequate' for use. This is despite substantial upgrade work on roads in the eastern seaboard. On the other hand, state roads vary greatly in quality and increased traffic is reducing local amenity. There has been no significant improvement in rural roads, despite the government's Roads to Recovery program of 1.2 billion Australian dollars over

the 4 years (Australian Infrastructure Report Card 2005). Table 6-2 shows that
the condition of road assets in Australia is far from being satisfactory.

Asset Type	Australia 2005	NSW 2003	VIC 2005	QLD 2004	WA 2005	SA 2005	TAS 2005	NT 2005	ACT 2005
Roads									
National	C+	C+	C	C+	B-	C	B	B-	-
State	C	C+	C-	C	B-	C-	C	C-	-
Local	C-	C-	C-	C	C+	D	D+	C-	-
Overall	C	C	C-	C	B-	C-	C	C	B

Table 6-2: Condition of Rail Infrastructure Assets in Australia
Source (Australian Infrastructure Report Card 2005)

Australia maintains one of the most extensive road networks, per capita, in
the world. According to Australian Bureau of Statistics (ABS 2005b) in year
2004 Australian road system comprised 810,641 kilometres, of which 336,962
kilometres (approximately 41.5 %) are bitumen or concrete roads. The same report states that New South Wales has the greatest length of bitumen or concrete
roads (91746 kilometres), representing just over half of all roads in that state, and
the Australian Capital Territory (ACT) has the highest percentage of total road
surface consisting of bitumen or concrete (95.3%), while the Northern Territory
has the lowest percentage of such roads (29.2%).

In recent past there has been increased interest in instituting strategic asset
management systems for roads, both at the government as well as industry level.
However, the results of the adoption of these systems have been quite varied. In
general, state government authorities have good amount of information and are
at an advanced stage in implementing these systems, for example, the New South
Wales's RTA annually publishes data on quality of roads and bridges (RTA
2000). However, despite the availability of requisite information, quality of roads
has not improved significantly. In response to this, Austroads, which represents
Australian and New Zealand's major road authorities, has developed a comprehensive performance evaluation system (Australian Infrastructure Report Card
2001). However, these systems only work where performance indicators are embedded into the asset management initiatives. For example, in an audit report of
national highways system (ANAO 2001) the Auditor General criticised inadequate asset management by the Department of Transport and Regional Services,
and noted that it could find no correlation between the performance indicators
and the agreed road conditions to be achieved by the States and the annual fund-

ing allocation to be provided by the Commonwealth (Australian Infrastructure Report Card 2001).

It has been argued that the lower than optimum condition of road assets is due to the lack of consistent and consolidated information and the multiplicity of asset management systems with no common grounds. This argument was highlighted in the BTE (2001) paper on local roads, which stated that no national figures are available on the physical conditions of local roads, on the numbers of vehicles that travel over them, or on the tonnage of freight that they carry. The Victorian Department of Infrastructure (2000) investigated these issues and recommended to recognize asset management as a corporate responsibility, rather than a technical responsibility. The study strongly argued the role of information in effective asset management, and emphasised the,

a. need for good quality information;
b. need for comprehensive asset management planning;
c. need for community involvement in establishing service standards;
d. need for rigour in financial assessments; and
e. need for performance measurement of asset management.

In general the national and state roads have been improving over the years, but rural and local roads have been a matter of concern. This concern is reflected in the results of three surveys which report, that

a. 62% of rural Councils indicated that their community would regard their roads as unsafe (Moree 2001);
b. 50% of motorists rated local roads as average or less than average (AAA Survey 2000); and
c. 67% of regional industries rated regional roads as average or less than average (AIG 2001).

6.3.2 Asset Management at Case B

Asset management at Case B is different than the usual asset managing businesses in a number of ways. The subject of the case, i.e. the construction companies participating in consortium, are going to own the asset (though these companies do not have the sole ownership and in fact it is the consortium as a whole that will have the ownership) for a specific time frame, after which it will be handed over to state government or its agencies. Nevertheless, during the timeframe of ownership maintenance and rehabilitation/refurbishment will be the responsibility of the consortium. A civil infrastructure asset has a large lifespan, which is generally spread over a few decades; therefore the consortium may not be involved in some of the activities of the asset lifecycle, such as asset rehabilitation/retirement/demolition.

6.3.3   Information Technologies at Case B

There are a variety of technologies being used at Case B projects, which range from simple Microsoft office applications to ERP systems. However, major technologies employed include, JD Edwards, Lotus notes, estimating package on buildsoft, CAD, and Optus inCITE (Out of these technologies the joint venture or the case organisations have taken up Optus inCITE platform solely for these BOOT projects). Optus inCITE is a construction industry trading exchange accessed through a web portal. The trading exchange came online at the beginning of May 2003, and the case organisations were involved in driving the launch of Optus inCITE. Both the organisations are committed to using this trading exchange throughout their operations and promoting its use in the Australian construction industry. Optus inCITE is owned and operated by Optus under a Trading Exchange Agreement with CITE Australia. CITE is an independent body in Australia and was formed by eight members of the construction industry for the purpose of implementing a neutral electronic trading exchange for the industry. To ensure that the exchange is developed in a manner that reflects the needs of the construction industry as a whole, an advisory board has been established. The board consists of representatives of different sectors of the construction industry and will continue to advise Optus of appropriate development of the trading exchange.

Optus inCITE provides the basic administrative functionality that is necessary to carryout construction projects. It provides document exchange services along with repository services so as to allow the companies participating in a project ready and easy access to project documentation, as well as to assist them in some of the essential construction workflow execution like design, purchasing, and tender/contract management through email, and document exchange and management services. Participating organisations have to pay an annual subscription fee to use these services. Although, this setup eliminates the costs of acquisition/ownership and maintenance of software, there are serious issues of capability, capacity, and technology involved. First of all, Optus inCITE does not provide an exhaustive set of construction management software services. In addition to this there are other issues like, lack of security and complete ownership of information; organisational adaptation to technology, which requires continuous training; and effort and finances required in integrating existing information systems infrastructure with Optus inCITE. Optus inCITE comprises three distinct applications (see figure 6-9) which are all accessed through an on-line portal (http://www.optusincite.com) by a single registered username and password. These applications are, document management; tender management; and purchasing.

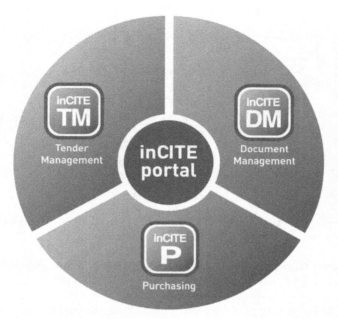

Figure 6-9: Optus inCITE Modules
Source: (Incite 2005)

The main objective of inCITE portal is to provide for automation of business processes, and in so doing allow for efficiency, reliability, and control of approval cycles involved in construction management. At the same time, with the ability to capture complete documentation of the project, it enhances accountability by allowing for traceability of documents. Theoretically the ability of capturing all information about a project and making it available throughout the company as well as throughout the project lifecycle allows for preserving and managing knowledge. This knowledge can thus be used and reused for better planning and execution of later projects. The document management module (figure 6-10) of Optus inCITE provides online management of documents, drawings, and communication. In addition to providing a central repository of documents it also provides for, automatic record of communications and changes to documents; version control; processes and workflows; and full traceability and audit trail.

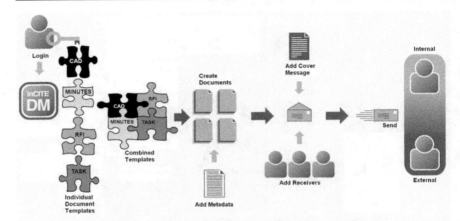

Figure 6-10: Optus inCITE Document Management Module
Source: (Incite 2005)

However, the biggest advantage that this module brings to a project is facilitating collaboration between different parties involved. Generally a construction
project is executed through a number of organisations that come together to
complete different aspects of the project. In such circumstances, communication,
ownership of information, and management of the same are extremely difficult.
In principle Optus inCITE provides a platform to resolve these issues. Tender
management module (figure 6-11) provides a multi-tiered approach to manage
the tender process by enabling flow of information from client to main contractor, as well as from main contractor to subcontractors.

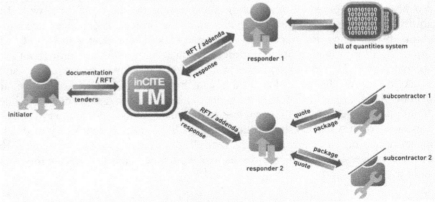

Figure 6-11: Optus inCITE Tender Management Module
Source: (Incite 2005)

It also allows for issuing tenders or subcontract/supply packages; submission of tenders or return of quotations; evaluation and comparison of tenders or quotations; management of addenda, and version control.The purchasing management module (figure 6-12) provides an online purchasing system and facilitates the issuing of request for quotation, approval processes, and management of individual company price lists and terms. This system can be interfaced with other organisational systems so as to allow for complete cost benefit analysis of a particular activity.

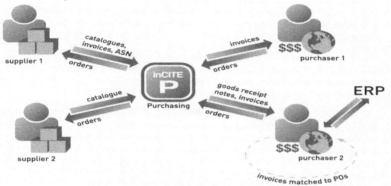

Figure 6-12: Optus inCITE Purchasing Management Module
Source: (Incite 2005)

6.3.4 Asset Design and Reliability Support

Asset design and supportability analysis for civil infrastructure assets is a little different than the industrial assets. This is primarily due to the fact that civil infrastructure assets have a very long lifespan and in certain cases it exceeds even 100 years. It is therefore almost impossible to make futuristic assumptions in terms of asset lifecycle support for the next hundred years with regard to costs, environmental impacts, and technology. Nevertheless, design considerations at Case B have to go through a number of reviews until all the stakeholders agree to it. This process is information intensive and involves a number of document exchanges, thereby prompting issues relating to version control, tracking of documents, as well as the changes made in them. Case B appears to be handling this issue pretty well with the availability of inCITE, which allows for a central repository of documents. In the words of a design engineer,

> It is great to be able to exchange real time information with 3-4 external companies anywhere and anytime. I don't always print out design and am quite happy to receive drawings in InCITE and reply in InCITE. I can track the design process quite easily.

Design Engineer (Case B)

The availability of design information in electronic format poses a significant advantage. Having this information available, in principle, should allow Case B to effectively administer construction of the asset, as well as use the same for future asset lifecycle decisions such as maintenance and refurbishment. However, this is not so due to the lack of integration of design, construction, and financial information. Both the case organisations are involved in constructions projects for anything between 1-5 years of time. End to end standardisation of technology, therefore, is considered as an undue investment. This is because, these things require substantial time, effort, and money and the collaboration has a limited time frame. Apart from this, each organisation is involved in a number of projects at any point in time and it is financially not feasible to invest in making their systems compatible with all of their business partners. This temporary nature of Case B impedes its participants from investing in information integration or interoperability with business partners' systems, which is argued by the IT manager of an organisation in Case B as

> One of the problems we have today is how do we integrate our different systems together, because we receive information that comes from many systems and companies. You may be excused to think that older technology is probably more difficult to integrate than newer technologies. I can tell you it is as difficult as it was before. We have struggled to achieve even basic integration. These projects are but two of the many projects that we have. I don't think it is cost effective to make changes in our technological base for the sake of these two projects.
>
> IT Manager (Case B)

Case B faces adhocracy, due to the nature of the BOOT project. Before commencing work, the participating organisations agreed on high level terms of running the consortium; however, businesses processes, their control and management was not given due consideration. As a result each organisation wants to push what suits its processes best. Consequently, the consortium lacks standardisation of practice. In an another example of adhocracy, quality of information is restricted to the drawings, since the same have been subjected to a number of reviews. However, the quality of the financial and administrative information cannot be guaranteed since it has not been audited and has travelled through a variety of different channels and systems. In the words of the civil reviewer at case B,

> We can ensure that the checks and balances that we can put in the systems are operating properly. But in terms of the type of information that gets entered, well, you can't check everything. You can check certain things that give a certain amount of assurance that things are OK.
>
> Civil Reviewer (Case B)

The survey conducted among the design staff employed at the two projects in Case B (figure 6-13) reveals a somewhat similar story . It shows a score of 3.04 for the technical capability of the information systems employed for design, which shows that designers are relatively satisfied with existing technology for managing asset design and associated documentation. It is for the same reason that information availability has been rated as 3.34. However, lack of training on new technology, i.e. Optus inCITE shows a slightly lower score for skills. Quality of information received an even lower score, which could be attributed to the ad-hoc nature of design information management, since it is not integrated with financial and administrative information.

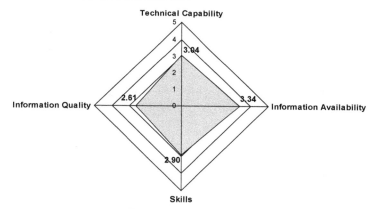

Figure 6-13: IS for Asset Design at Case B

6.3.5   Operational Efficiency

Case B's projects represent a networked organisation that comes together to execute specific projects. Consequently it is relatively difficult to enforce standardisation of practice and technology, since each participating organisation has its own set of policies, technologies, process workflows, and skills. Optus inCITE enables the organisations in Case B to communicate better and with relatively easy access to information. In addition, it was anticipated that the workflow processes developed and learning from the tunnel project could be applied to the road construction project, since the later started 18 months after the first project. However, that has not been the case; due to the difference in contractual and project arrangements, and additional parties involved in important processes, such as design review process. At the same time, lessons learnt at the tunnel project with regard to technology mapping were not used at the second project either. For example, there was a greater willingness to accept Optus InCITE at the tunnel project but the use of the same technology at the second project appears to be challenged a lot more by project staff due to the slowness, lack of intuitiveness and lack of availability of exhaustive construction processes on Optus InCITE.

Both the case organisations at Case B have been using Lotus Notes and have also developed a certain level of maturity in mapping its workflow through the application. However, in these projects they have to use Optus InCITE and its embedded workflow. As a result, staff at both organisations has been unable to reconcile with the changed software environment and require extensive training.

> Now all of a sudden we require too much IT knowledge and/or processes to operate effectively.
>
> Site Co-ordinator (Case B)

In terms of the process outputs, there is a mixed response to the satisfaction of staff with the use of technology. The slow processing speed and lack of integration with organisational systems, particularly the ERP system, has added to frustration of staff.

> Day to day activities like messaging and email and locating documents are taking much longer than they used to. The system is not responsive enough, not intuitive enough and not reliable enough resulting in a lot of frustration and negative feeling.
>
> Project Engineer (Case B)

> When you record all electronic exchanges it improves stakeholder accountability. Everybody has access to latest information, which reduces the chance of errors on site.
>
> Civil Reviewer (Case B)

The survey conducted among the staff at Case B reveals a mixed response. Figure 6-14 summarises the results.

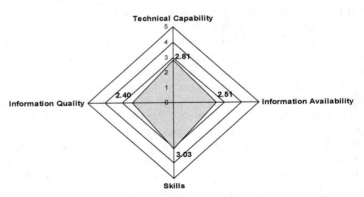

Figure 6-14: IS for EAM Operational Efficiency at Case B

The score of 3.03 suggests that staff are generally satisfied with the use of technology. This could be due to InCITE or the fact that technology is more or less the same at both the projects. The score of 2.81 for the technical capability shows that staff are aware of the capability of technologies employed at the project. However, issues with the effective integration and use of technology have had a negative impact on the overall benefits of technology, which has resulted in low score for information quality and availability.

6.3.6 Competitiveness

The idea of a BOOT project may appear cost effective in the sense that infrastructure asset are financed, built, and managed by commercial parties with the role of the government as a moderator, and eventually the asset being transferred to government any way. However, management of infrastructure assets require long term commitments and planning; whereas work practices in Case B have an ad-hoc demeanour and decision makers have their focus on the immediate short term. At the same time while the concept of BOOT effectively shifts risks, such as construction cost, time overruns, design faults, and operating cost overruns; yet the effectiveness of information systems and process execution is compromised due to lack of management and control of technology. For example, Case B has adopted inCITE as the main technology so as to provide standardisation across the consortium, whereas in actual effect information contained in inCITE lacks quality and participating organisations have found it difficult to or are not interested in integrating their information systems with inCITE. Lacking information quality has also affected another expectation from inCITE, that is of knowledge management. The consortium was unable to transfer significant meaningful and action oriented learnings from first project to the second. However, technologies like InCITE have the potential to provide strategic advantage through its potential for global cooperation and enhancing organisational image. This advantage, however, is dependent on management's ability to instil necessary cultural change to embrace technology as well as changes in the business processes. The chief engineer on tunnel project summarises management's commitment in this regards as,

> From the outset, when the decision was made for InCITE as the main tool, then its application should have happened right from that particular moment. This project started in Sept 04 and we are still (May 2006) umming and ooing about InCITE. We should sit down and work through all the cobwebs, recruitment issues, training, and a smooth transition to use this tool. My InCITE training was left up to me to book in and when you hear so many negative things about it. Its not something you rush to do.
>
> Chief Engineer (Case B)

The figure 6-15 shows Case B's management's perception of the role of information systems in enhancing the competitiveness of asset management. It is evident that technology is viewed as relatively well matured and capable of delivering the goods for strategic asset management. However, management does not have great deal of confidence in the skills of the employees, and it appears that using different software systems, i.e. JD Edwards, InCITE, Lotus Notes, and a range of other specialised systems is demanding a bit too much from staff. In addition, lack of integration of financial information with operational information and the issues with inCITE contribute to the lack of trust in information, which not only impacts the quality of decision making but also affects the management of asset lifecycle as a result of these decisions.

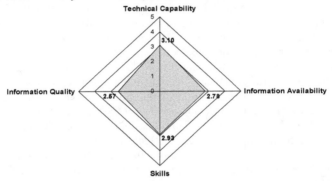

Figure 6-15: IS for Competitiveness of Asset Management at Case B

6.3.7   Evaluation Initiatives

Case B does not have a formal information systems evaluation strategy. Most of the technologies employed at Case B are actually decided by the consortium and are to be used by the participants regardless. However, functional managers and site coordinators have done some pre implementation evaluations to assess the usefulness of technology to their functional areas. Nevertheless, they commutate the results of these exercises to top management and suggest solutions as preferred choices of technologies to manage projects. It is then up to the top management if they would approve the use of technologies or discard the idea or suggestion. These evaluations do not make any input into strategic decision making for asset lifecycle management technology infrastructure.

6.4   Rail Infrastructure Asset Management – Case C

Company C is a rail transport organisation based in one of the largest states of Australia. It owns and manages 3.8 billion Australian dollars worth of rail network that stretches throughout the important industrial and agriculture stretch of Australia. The company has been in operation for 141 years and is one of Australia's largest passenger, coal, and freight transport provider with annual revenue of 2.5 billion Australian dollars per year. In the financial year 2005-06, more than 2600 staff of Company C operated approximately 260000 passenger services, and carried over 54 million passengers. This was an increase of approximately 9% over the previous financial year. In all Company C employs more than 13000 staff and provides a broad range of freight services to a wide customer base in many industries in Australia. Company C manages more than 5000 services every month, through its 9500 km rail network. It has presence in most major Australian cities and has access to more than 50 terminals and depots spread all over Australia, which enables the organisation to provide a one stop shop solution to industries and other freight customers. In so doing, Company C enables vital supply chain management support in terms of transportation, warehousing, shipping and port operations and distribution.

Company C's parent entity's pre-tax profit were 89.2 million Australian dollars for the financial year 2005-06 and for the same period the company's pre-tax profit was 93.5 million Australian dollars. Company C's sales revenue exceeded 2 billion Australian dollars mark in the year 2005-06 for the sixth consecutive year. Company C's state based fleet includes over 12300 units of rolling stock, which includes more than 10200 wagons, 508 diesel and electric locomotives, 143 three-car electric trains, and 177 passenger carriages. All of which are used to transport people, coal, bulk or containerised freight. Company C employs in excess of 1400 staff (including 176 apprentices and trainees) at four geographically dispersed locations all over the state to manage these units. In the year 2005-06 Company C overhauled or maintained more than 3000 of these units. Being a publicly owned large organisation, Company C has a fairly large spared of management functions, as illustrated in figure 6-16.

Asset lifecycle management activities at Company C are spread over the Network Access, Infrastructure Services, Rolling stock and Component Services groups. In addition to providing track access, Network Access also manages access to corridors, major yards and telecommunications services. Whereas the Infrastructure Services Group constructs, maintains and manages Company C's rail infrastructure to deliver a safe and reliable network. Rolling stock and Component Services group manufactures and overhauls rolling stock (including locomotives, carriages and wagons) for heritage, national, as well as city fleets.

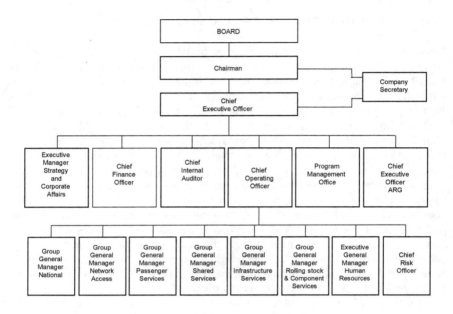

Figure 6-16: Company C's Organisational Structure

6.4.1 Australian Rail Industry

Rail infrastructure is a vital component of Australia's national transport infrastructure. Australian railway industry has an annual turnover in excess of 8 billion Australian dollars and employs approximately 75000 staff (Austrade 2006). However, till 1990's railways in Australia was publicly owned and had vertically integrated operations. Concerns for improving efficiency necessitated reforms, and thus the industry underwent major changes through which some state railways transformed into private/public structures, separating ownership, operation and regulation (Australian Infrastructure Report Card 2001). The existing rail infrastructure in Australia can be categorised as, urban passenger services; regional / interstate passenger services; regional and interstate freight services; and bulk materials freight, e.g. coal, grain, sugar, minerals.

Australian rail network is dominated by three railway gauges, i.e., narrow, standard, and broad gauge. Approximately 14831 (2462 electrified) route kilometres are laid to the narrow gauge, around 28662 (1397 electrified) route kilometres are laid to the standard gauge, and around 4015 kilometres are laid to the broad gauge; and approximately 230 route kilometres are laid to dual gauge (CIA 2006). Out of this network, urban railway infrastructure in Australia is primarily used for the passenger train services; however major lines are shared with freight services. On the other hand the major portion of regional and inter-state rail infrastructure is used for freight services alone, whereas, a mix of re-gional and interstate infrastructure is used for bulk material freight. Most of the rail infrastructure operating companies are in the private sector and are profitable enterprises that operate in highly competitive domestic and international mar-kets. With regard to international market, Australian railways industry has made significant impact in established markets of South America and South East Asia along with major project wins across the globe such as track construction associ-ated with the Taiwan High Speed JV; track construction, Automatic Fare Collec-tion, signalling and passenger vehicle maintenance projects in Hong Kong; and simulation technology into the UK (Austrade 2006).

Funding from federal and state government in road infrastructure exceeds investments made in the rail network, which affects the ability of railway organi-sations to compete with road freight. At the same time, it lays extra pressure on management of asset infrastructure. However, from the investments made in the rail infrastructure, metropolitan networks have traditionally been under funded, bulk freight, such as coal lines have been appropriately funded and thus are per-forming quite well. Table 6-3 shows the state of rail asset infrastructure in Aus-tralia.

Asset Type	Australia 2005	NSW 2003	VIC 2005	QLD 2004	WA 2005	SA 2005	TAS 2005	NT 2005
Rail	C-	D	C-	C+	C+	C	-	A

Table 6-3: Condition of Rail Infrastructure Assets in Australia
Source (Australian Infrastructure Report Card 2005)

According to Australian Rail Track Corporation (2005), fixed and deferred assets in Australian Railway industry totalled 35.2 billion Australian dollars in 2004-05 of which 23.3 billion Australian dollars or 66.2% were attributable to passenger services. For passenger operators, 36.6% of the value of fixed and deferred assets was attributable to land and buildings followed by track infra-structure (31.6%) and rolling stock (22.3%). With regard to freight operations, 60.9% of the value of fixed and deferred assets was directed towards track infra-structure, followed by rolling stock (20.0%).

6.4.2   Asset Management at Company C

Company C manages a number of assets, including buildings, such as godowns; rolling stock, and track assets. In the state infrastructure report card 2004, Com-pany C's assets were rated at three levels, i.e. metropolitan network for which it was given a grade of C, due to high demand of service and low investments in asset renewal and refurbishment; freight networks, received a B+, due to the investment from coal industry in maintaining the assets infrastructure; other networks, which received a D+ due to inadequate funding and poor service provision.

Company C has generally followed a whole lifecycle approach to managing its assets. In order to optimise the assets, the company has developed an asset management framework that has a ten years time span. However, it is subjected to minor modifications so as to make it relevant to current legislations as well as changes in the market. The asset management framework is based on five key elements, which are

a.   Company C's network development plan;
b.   Company C's network maintenance plan;
c.   an alliance-style maintenance and project agreements with specified goals;
d.   a detailed performance monitoring framework; and
e.   independent asset condition and service provision auditing.

Company C's asset management framework also includes a financial asset corridor model to provide historical and projected indications of the financial performance of each asset in company's rail network. This model accounts for the revenues, capital investments, maintenance activities, capital charges and internal costs and service charges. Company C has a number of systems aimed at enhancing and maintaining asset management capability, which encompass range of asset lifecycle management dimensions. These systems include, procurement and materials logistics; track and structures performance management; detailed long and short-term planning advice; rail infrastructure condition monitoring; asset inspection and safety auditing; regulatory and operational compliance assurance; and property and contract management.

6.4.3   Information Technologies at Company C

One of the core aspects of the proposed asset information management framework is IT reform, which is an initiative to reduce the number of systems operating within Company C. The company expects to save 20 million Australian dollars in sustainable cost savings from standardisation of foundation information systems and by increasing the visibility of all spending on IT. Company C is also participating in two federal government funded cooperative research centres to enhance its IT platforms. The scope of these technologies ranges from standalone process facilitating tools to integrated decision support and planning systems. It is from the same platform that the company has recently set up a scheduling optimisation tool to increase the speed and effectiveness of train, crew and maintenance scheduling on track, and its business intelligence technical infrastructure. However, major technologies employed by Company C are SAP R/3; CAD; CMMS; and a variety of industry specific asset lifecycle management softwares such as RailFrame, TRIM, PST, V0, RIMS, and RDMS. Company C appears to be quite proactive in trying out new technologies and have thus formed a pretty robust technological base for some of the asset management functions. Case C's IT manager summarises technology adoption approach as,

> We are not early adopters, and we are not explorers and we are not easily influenced or driven by whatever the latest thing on the market is. It is needs driven and business case driven. Basically in past our motivating factors have been tactical needs of individual areas, so it hasn't been strategic at all but its moving towards being more strategic mainly for information integration. We now have stronger governance and cost focus, since we are now viewing ourselves as a market player as we are expanding nationally and are moving into more commercial roles.

> IT Manager (Company C)

6.4.4 Asset Design and Reliability Support

There are a variety of assets maintained by Company C, such as carriages, godowns, railway track, bridges, signals, and engines. However, for this case study the subject under study is the below train assets, i.e. track, bridges, and other civil infrastructure. In this regard, Company C has an old set of asset infrastructure as majority of these assets were laid in 1920s and 30s, with some even earlier. However, with changes in technology these assets have been continuously upgraded, for example for high speed and heavy rail application the old gauge doesn't fit and therefore has been upgraded to meet the needs of faster, and reliable technology.

In the design/redesign function, Company C utilises a range of technologies to aid operational and administrative workflow. However, the technologies used for design are micro stations, AutoCAD, and civil design software 12 D. Asset design/redesign is primarily governed by what the Company C terms as 'community benefit', which actually is expansion of the span of control and thus could be termed as expansion of market share. However, since it is a semi government organisation its assets' need assessment is in response to industrial freight customers as well as passengers utilising its metropolitan and interstate services. As mentioned before, the other major impetus of asset construction, renewal, or rehabilitation comes from the technology refresh. There is not too much of complexity involved in designing track assets and therefore the process is fairly stable. According to a design engineer at Company C,

> A piece of track looks the same today, looked the same five years ago, and will look the same in five years from now. However, it's the formation that keeps on changing............Although we have got fair bit of say over what software applications we use, we miss the old system where we had somebody that was sort of monitoring what was happening in the market with regard to design software

from across Company C because at the moment where I see some degree of connectivity with civil engineering design, there is little connectivity when we go across other areas, e.g. electrical design.

Design Engineer (Company C)

While designing assets, design engineers do take into consideration the asset workload and work out the need profile. However, it is all done manually or with support from simple Microsoft Excel based spreadsheets. Traditionally, asset design followed an established process, where the design engineers surveyed the area and identified particular routs. They would then design the asset accordingly. There has also been heavy reliance on the knowledge of field staff in designing or refurbishing sections of asset, since they are the closest to the asset and are in a position to predict some of the issues that the asset may be exposed to. Nevertheless, the design information is held locally in the regional offices, apart from the original drawings. This information is not exchanged between regional offices or with other functions of asset lifecycle management. In addition, recommendations on asset lifecycle supportability design form a part of the design feasibility study, however the actual information remains with the designers and is not exchanged or transferred to a system where it could be reused by other functions or departments. The business development manager of Company C states that,

We know we should capture, management, and maintenance of knowledge for future generation of Company C, so we don't have to reinvent the wheel every time. We are long away from that. In terms of information we have proliferation of tactically disparate databases and spreadsheets. We have got the information but it stays with designers. It is not exchanged and even if it were exchanged it could not be merged with other information.

Business Development Manager (Company C)

In the survey conducted among the designers at Company C (figure 6-17), respondents indicated a below average confidence level on the suitability of technology to execute their work. At the same time they showed an enhanced level of confidence in their skills to use these technologies. However, they showed a modest decrease in the availability and quality of information as compared with the capability of technology to execute design related asset management processes. This is mainly due to the inability of the organisation to store design and related information. Company C has a vast resource of scanned copies of design drawings; however, in the absence of associated metadata this information is of little use.

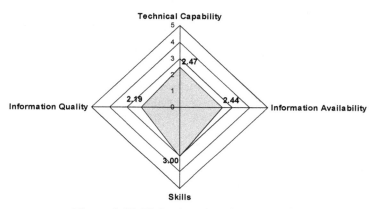

Figure 6-17: IS for Asset Design at Case C

6.4.5   Asset Operation Perspective

Company C caters for metropolitan as well as countryside track assets, and therefore is not only concerned with the traffic on the tracks but also the weight borne by these tracks over the period of time. Traffic is managed by state of the art software that manages as well as allocates traffic on the tracks; whereas, the condition of the track is monitored through sensors and is also inspected manually. Human inspections actually constitute majority of condition monitoring due to the remoteness of the track assets. Company C has an extensive network of track inspectors, which includes a substantial number of indigenous Australians who are well known for their knowledge and familiarity with outback terrain and geography. Company C relies heavily on their tacit knowledge, and these track inspectors have also proven to be extremely reliable sources of track information. However, there has been no effort made to record information collected through these manual inspections. Whereas, there are certain aspects of asset operation that seem to be over automated, as is evident from the response of operations manager,

> For a case of a broken rail, essentially it's about train coming off. One system records broken rail, which goes to the network controller who can stop trains from going on the track. Another system records the same incident the same information in a track incidence system to raise signal alarms. Yet another one of the systems records the same incident in the rail defect management system, such that a request could be generated to fix it. Now you have the same information available in three different systems. There is not only duplication, in fact triplication of information. Information in each system has a bias towards a particular function, so which version is more credible?
>
> Operations Manager (Company C)

This symbolises the typical behaviour of a public sector large organisation. In this case, each function trusts its own information and does not believe in sharing the same with other functions. As a result, there is significant wastage of effort and finances, and quality of information is undermined due to information redundancy and lack of integration at various levels. Even though there is significant automation of processes, information systems in asset operation at Company C are far from being productive. The same story is illustrated by the survey conducted within employees associated with asset design (figure 6-18).

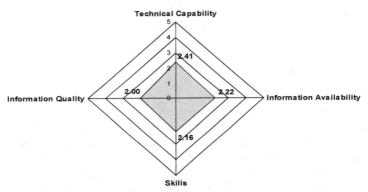

Figure 6-18: IS for Asset Operation Support at Case C

Multiplicity of information, lack of its quality, and inability of the organisation to capture information from manual inspections, are some of the issues that impede effective use of technology. Staff attribute this to the inaptitude of technology. The survey results reflect this through a low level of trust in technology and the information contained and processed systems.

6.4.6   Asset Maintenance and Lifecycle Support

All maintenance in Company C is carried out in house, and no part is sublet to a third party. It follows a periodic preventive maintenance schedule and since the company maintains a number of different assets this schedule varies for each type of assets. Though track assets are fairly stable and do not develop failure conditions too frequently, the inspection of track assets is held frequently. Information on condition of an asset as well as the treatments carried out are kept with the regional offices and only a summarised version of this information (chiefly financial) is communicated to the corporate head offices, unless the track requires a major overhaul or relaying. Major software tools used in maintenance function are the Rail Infrastructure Maintenance System, and Royal Defects Management System. These systems help in condition monitoring, defect detection, and maintenance scheduling and execution; however these systems are not integrated. A possible reason for disparate technical infrastructure may be the propensity of engineers to concentrate on getting the immediate job done. However, little consideration has been given to getting the same job done more efficiently and cost effectively. Therefore, although more or less each activity has a separate information system, but the information thus captured cannot be used for any strategic advantage.

> Where it is (failure), what type is it, what do we need for it, yes we have the systems that help us with these activities........but these technologies are stand-alone. There are PDAs, there is GPS, there are all sorts of technologies. What we needs need is a single view, which helps us in finding out how much did we spend on our assets, what expertise was detailed, what spares were required.
>
> Maintenance Manager (Company C)

Maintenance function conforms to a risk assessment standard for safety management (which has been developed by Company C), which covers safety, environmental, commercial, human resource, financial aspects and assessment of technology within this context. However, this technological assessment concerns hardware assessment from an OH&S (Occupational Health and Safety) point of view only and does not audit information or information systems. Consequently, pools of data (with varying degree of quality) can be found throughout the organisation that cannot be put to effective use, because nobody knows what detail of information is contained in them.The survey conducted in the maintenance function Company C (figure 6-19) indicates a relatively higher level of satisfaction with technology and information acquisition than other functions in Company C.

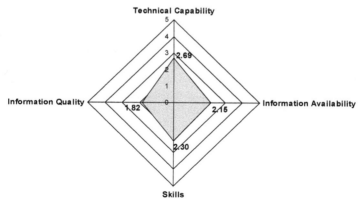

Figure 6-19: IS for Asset Lifecycle Maintenance and Support at Case C

However, the score of 1.82 for the quality of information can be attributed to lack of integration of organisational information systems infrastructure, as well as immaturity of technology in the organisation.

6.4.7   Operational Efficiency

As a semi government organisation Company C characterises most of the advantages and disadvantages associated with a hierarchical structure. There is little cross functional and cross departmental collaboration with each function working within well defined boundaries. Consequently, the general approach is 'if it ain't broke don't fix it'. The business development manager provided some insights into the organisational culture by stating that his office is at the same floor as many of the electrical engineers, but they have never spoken to each other. This function centred approach has translated into the way information systems are utilised in the organisation. There are a range of different systems and each aiming to accomplish individual tasks, but they are not all integrated to provide a consolidated view of information at a higher level in the hierarchy. In the words of the Network Access Manager,

> There is huge range of standalone information collection devices, which primarily collect historical information. So it's range of historic information that's available to us. What we want to do is to actually get all of it to be available at one spot, get all of the systems talking to each other, reduce the duplication of data so that when we go in and ask for any query we know it's the same. Then there is standardisation of data across the organisation. We want to move beyond the individual data management to predictive issue based management.
>
> Network Access Manager (Company C)

Heavy reliance on tacit knowledge and the inability of the organisation to preserve this knowledge poses a significant challenge to creating a learning culture in the organisation. The lack of standardisation of data and practice is the major impediment in enabling organisational learning; however it is not the only cause. To sieve out learnings from the execution of routine business, integration and interoperability of information is as important and facilitative as developing the culture of information sharing and exchange to achieve higher levels of coordination and cooperation. In Company C there is hardly any information exchange between asset lifecycle stages, which prevents the realisation of an information driven integrated view of asset lifecycle. However, management in Company C is aware of this issue,

> When we talk about the big picture, you may have one piece of information and someone else can have the other piece. He doesn't necessarily see the other piece of information which together can actually point you to a totally new area. For continuous improvement we have to change technology and also have to change the way we do daily business.
>
> Infrastructure Group Manager (Company C)

The survey conducted among the asset management stakeholders in Company C (figure 6-20) indicates a score of 2.13 for the technical capability, which is largely due to the fact that information systems have not lived up to their true potential in integrating the organisation, enabling collaboration, and providing the integrated view of the asset lifecycle for better execution and management of assets.

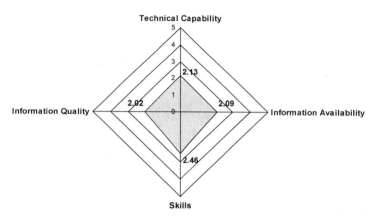

Figure 6-20: IS for EAM Operational Efficiency at Case C

Although these issues are not entirely technical and have a lot to do with change management and organisational behaviour; yet in the perception of staff these are issues originating from the way information systems are deployed in the organisation and the inability of technology.

6.4.8   Competitiveness

Being a state owned organisation, Company C has traditionally been insulated from competition. However, with deregulation their business environment is changing and the company is expanding its operations to other geographic location in Australia. At the same time, with programs like Auslink (Federal government's initiative to improve roads) Company C is facing increased competition from alternative service providers. Nevertheless, it has been only recently that the top management has started considering itself as a market player rather than a monopoly. This change is quite evident in the way Company C is managing its information systems. Traditionally, these information systems have a narrow focus and are primarily being used for operational support with little or no planning, management, and strategic support. However, this is changing fast,

> We are going beyond total (asset) life (profiling), we are going to total community benefit and trying to financially quantify some of those things such as en-

hanced access stations, and the sort of benefits of integrated bus-train interchange to the community.

Business Manager (Company C)

At Company C information systems infrastructure is disjointed and includes
many ad-hoc solutions. Instead of using core information systems technology of
the organisation, different functions prefer to use simple spreadsheet and database applications. Their use is justified as 'they are easy to use', and that 'they
can be customised to meet changing needs'. Thus data that is potentially useful
for other functions or processes remains hidden in these ad-hoc solutions. This
lack of control and standardisation has resulted in islands of useful data in the
organisation without being put to effective use. The risk officer summarises the
situation as,

> For asset life cycle decision support we generally rely on historic data. There is
> not a huge amount of data available though. It (decision making process) is a lot
> based on engineering knowledge, lot of our people have been involved in opera-
> tional management of the assets. So they know how the asset performs and be-
> haves. They know the discreet life cycle of the asset components, and by putting
> those things together we can come up with the forward projection of asset.
> There is no rocket science there, its based on personal knowledge of particular
> engineers involved.

Risk Management Officer (Company C)

Company C is using SAP as the core information systems for asset management. SAP was implemented in the organisation from finance/human resource perspective and was later adopted as a core asset management information system. Even though the asset lifecycle management functions in the organisation are still struggling with the basic questions whether the technology has the
depth of detail and elegance required to manage assets. In the words of Group
Manager Infrastructure services,

> With respect to asset management we are using SAP to store all data. It doesn't
> provide engineering state of life cycle, since the data is not integrated. For ex-
> ample, we know how much are we spending on track maintenance, but we can-
> not straight away find out how much was spent where. This information is not
> integrated with maintenance or design or operation. We are in the process of
> building some systems now and our group is also reviewing several different
> life cycle scenarios, costing and planning tools for our track. But at this point,
> no we haven't got an integrated life cycle asset management.

Group Manager Infrastructure Services (Company C)

In line with the responses from the senior management in Case C, the survey
results of the management's perception about the effectiveness of information
systems for competitiveness of asset management strategy reveal the same story
(figure 6-21). Low scores on all accounts suggest the uncertain and tentative

attitude towards information systems for executive decision support, and the inability of existing information systems to make any strategic advisory contribution.

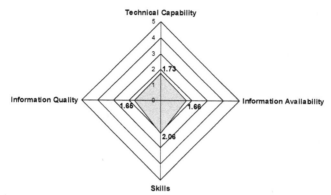

Figure 6-21: IS for Competitiveness of Asset Management at Case C

6.4.9   Evaluation Initiatives

Company C does not have a formal information systems evaluation system or
procedure in place, which explains the proliferation of the enormous number of
custom made spreadsheets and databases. However, there is an informal process
of evaluating technology is some of the departments, but this evaluation is by no
means comprehensive and basically evaluates engineering aspects with regard to
technology use. The IT manager of company C stated that,

> We do a bit of feasibility evaluation but that takes only a superficial view at
> some of the scenarios, but we don't do post implementation reviews at all. This
> evaluation (feasibility) is by no means comprehensive. Once the project kicks
> off you find that there is a lot of information, a lot of things that you should
> have given a little bit more thought to. Implementation of technology costs a lot
> of money and I think you have to really invest a little bit upfront to determine
> whether the chosen solution will give you the value that you are looking for be-
> cause vendors can make everything look pretty.
>
> IT Manager (Company C)

Company C has recently concluded an evaluation exercise, whose purpose
was reorientation of asset management plan. As a part of this evaluation, information systems for asset management were also considered. However, this consideration was limited to what information systems might be useful, rather than
assessing the contribution or suitability of existing information systems to asset
lifecycle management processes. Nevertheless, there is an encouraging sign that

these evaluations have the support of top management, as stated by the IT manager of Company C,

> Evaluations are certainly not frequent but we hope to have them on ongoing basis. Any resulting IT reforms are reported directly to the chief operating officer, the second most powerful person in the organization, and he is really serious about getting better value out of our IT systems and stopping this growth of temporary tactical solutions. We are a long way from achieving significant benefits though. In order for these evaluations to work for asset management, first of all we must agree yes there is a problem, yes we need to sit down and have a really good look at this, yes we are committed to solving it and that's going to take a lot of work.

> IT Manager (Company C)

6.5   Summary

This chapter presented the empirical study conducted for this research. Each case has been explained in detail, and included discussion on asset management initiatives within the industry that the case organisation operates in, information systems infrastructure, and the use of information systems in each stage of asset lifecycle management. Direct quotations from the case study interviews have been used to complement the overall point of view being presented. In addition, survey results have also been interpreted to help develop the overall picture of information systems maturity in asset management processes with each organisation.

The next chapter provides analysis of these case studies to bring out common themes and concepts. Both within and cross case analysis will be carried out to uncover themes that will form the basis for answering the research question set forth at the start of this research.

7   Research Discussion

This research addresses the principal research question, 'why do asset managing organisations generally fail to evaluate the performance of information system based asset lifecycle processes, which could enable them to better understand and manage the performance and needs of asset lifecycle?' However, addressing this question requires resolving, 'how do information systems facilitate alignment of strategic asset management considerations with overall business strategy and organisational design; what factors impact institutionalisation of information systems based engineering asset management processes and their performance evaluation; and how information systems based asset lifecycle management processes should be evaluated?'. This chapter is based on the discussion in previous chapters and attempts to answer these questions by taking into consideration the findings from the case studies and the literature review conducted for this research.

The chapter begins with presentation of findings from each case with regard to the premise of this research. This is followed by cross case analysis, which brings out the common themes from the three cases to reveal the issues posed to information systems based asset management and information systems based asset management evaluation, and factors that impact information systems implementation for asset management. A detailed discussion of these themes leads into justifications and explanations of the answers to research question of this research.

On the basis of the evidence presented in this chapter, it is particularly argued that implementation of information systems for asset management dynamic process and has a strong social orientation, and their evaluation is action oriented learning exercise. The chapter develops an information systems based asset management organisational alignment framework, and information systems based asset management evaluation framework. The outputs of this chapter could, thus, be used as an implementation guide for information systems for asset management, information systems based asset management governance framework, and integration guide for information systems utilised for asset management.

7.1   Case Studies Analysis

This section presents findings from each case and highlights the learnings with regard to the research scope of this research. The discussion on each case is followed by cross case analysis that summarises the overall findings and highlights common themes.

7.1.1   Water Infrastructure Asset Management – Case A

Company A is a representative case of semi government public sector organisations. It has a hierarchical structure, bureaucratic culture, and centralised decision making. As it is evident from the case study, Company A does not have a standard policy to information systems implementation throughout the organisation. It does not audit the performance of its asset management related technical infrastructure, which could highlight the information and technology needs of asset management processes.

Company A does not conform to a specific enterprise architecture or information model. As a result, their technological infrastructure is based on ad hoc assumptions and is increased and developed on as required basis. There is little consideration given to the pull factor of information requirements of asset management processes, which defines what technological and organisational infrastructure, and human skills are required for efficient execution of the processes. Therefore, technology pull, in theory, facilitates choosing appropriate technology that the organisation needs to adapt to and initiating an appropriate change management strategy. However, the general technology implementation philosophy being followed in Company A is based on technology push. In this case, technology is pushed into the organisation (as a result of the mimetic isomorphic pressure of the institutionalisation theory) and then the whole organisation has to adapt to technology. This is evident from the fact that the SAP was implemented due to the pressures from regulatory agencies, rather than as a response to fulfil the information needs of asset management processes. In this case, not only that there is a major disconnect between the process and organisational needs, profile of technology and process maturity, but Company A also finds it difficult to adapt to new technologies and integrate them with the existing technical infrastructure of the organisation. Thus, at the heart of their technology implementation strategy, Company A regards investments in IT infrastructure as a support activity with little strategic value.

Company A, being a public sector organisation, has traditionally been immune to competition, due to which it is focused on internal efficiency without much consideration to organisational responsiveness to competitive pressures. Its principal aim in using information system is to seek and achieve business process efficiency. However, it does so at the peril of being responsive to business environment to be competitive, because its information systems are not designed and developed to provide strategic support. Company A has only recently expanded the scope of technology to integrate with business partners, by investing in technical infrastructure to integrate with customers and business partners.

Technology implementation at Company A conforms to a deterministic approach and is aimed more at achieving perceived benefits from technology rather than the cause and effect relationships that enable these benefits. The lack of vision and the unavailability of enterprise wide information systems architecture have led to the emergence and utilisation of plethora of ad hoc solutions throughout the organisation. These solutions symbolise a number of small organisations within company A, and are specific to departments or functions, are isolated from the mainstream technological infrastructure, serve the needs of specific processes, and have a narrow focus. As a result, each asset lifecycle function is focused within its own boundaries and there is less interaction with other functions. The information collected and processed by each asset lifecycle function is aimed at fulfilling its own demands rather than contributing to the overall objectives of asset lifecycle management. Consequently, the available information is biased, not credible, incomplete, and is of little use outside the functional boundaries from where the information originates.

Since there is no collaboration or cohesion between different functions and departments, information systems are being utilised for simple process automation and record keeping, and not for any strategic advantage. At the same time, inability of the organisation to integrate financial information with operational information does not provide a complete picture of asset lifecycle efficiency. This inability hampers the ability of the organisation to create an integrated information enabled view of asset lifecycle, and impedes holistic decision making for effective asset lifecycle management. There is no asset proofing done at any level and decision making regarding various aspects of assets' lifecycle is carried out by individual functions and is not all inclusive or comprehensive. In these circumstances, it is no surprise that the organisation does not evaluate the performance of information systems based asset lifecycle management processes. Even if it did, there are all the more chances that the results would lack credibility and validity.

Top management at Company A is not technology savvy and the functional staff considers technology utilisation as an unnecessary addition to their routine jobs. It is well summarised in the knockout quote from their group manager who stated that "some would argue that we are in an asset based industry and not an intellectual property based industry or anything like that". It shows that top management is not aware of the capabilities of technology, which would allow them to create a vision or a technology roadmap that could couple organisational needs with technological capabilities. Same reason can be stated for lack of commitment and resources to train/up-skill functional staff on existing information systems.

Company A does not seem to be interested in preserving knowledge. Even though there is heavy reliance on tacit knowledge in execution of various asset management processes. The management of explicit knowledge also seems quite difficult, since there is plethora of paper based as well as electronic information resources throughout the organisation. The way in which organisational technical infrastructure has evolved does not suggest that the organisation is making a concerted effort towards preserving its intellectual capital. At the same time there are no efforts being made to convert whatever information is available into useful knowledge through data mining or business intelligence tools and techniques. It is worth mentioning that the organisation is considering investing in a suit of business intelligence tools and applications and is hoping to improve its information resource through reverse engineering.

Asset management and enabling technology has different ownerships within the organisation. Assets are owned by engineering services and IT function (excluding operational technologies) is owned by CIO and IT department. Within asset management there is multiplicity of ownership, i.e. asset lifecycle management is under engineering services and maintenance is owned by engineering maintenance department. Engineering services has no control over execution and management of maintenance. This has resulted in multiplicity of controls that often come in conflict with each other. However, in terms of information systems it has meant a clear distinction between the way information and operational technologies are procured, maintained, and managed. Consequently, there is fragmentation of technological landscape with isolated data sources that lack quality, usability, and reliability. On another level, these issues have resulted in incompatible technologies that have further complicated the technical infrastructure of the organisation.

It is evident that there is no culture of ex-ante or ex-post evaluation to measure the effectiveness of technology in achieving the aims and objectives that the organisational stakeholders associate with it. Recently Company A underwent an audit of existing asset management strategic plan, which was carried out by external consultants and was done to fulfil regulatory requirements. However, the exercise was theoretical and does not appear to be all encompassing. For example, technology was also evaluated in the same audit, yet the focus was at a higher level and dealt with concerns such as, what technologies are suitable for asset management areas in general. In principle, employing external consultants has a major advantage in the sense that they are not biased. However, the assessments made by consultants were not plausible for some issues. To begin with this evaluation was aimed at what technology 'might' help in achieving the objectives of the process. Furthermore, the evaluation team lacked representation of functional staff that participate in the execution of routine business processes; are closest to information sources; or actually capture, process, and use this information. Consequently the results of the asset management audit did not account for the operational issues encountered by the organisation. At the planning and strategic levels, there was no benchmark standard against which performance was measured and the evaluations did not enable any actionable learnings. In crux, the technology evaluation in this exercise represented a wish list of what technology might be required to enable asset management processes, rather than providing a need assessment of the existing processes (which truly reflect the organisational needs, since the organisation has evolved them over a long period of time) in terms of information systems.

In summary, there is a philosophical difference between the character of the organisation and information technologies. As a public sector hierarchical organisation, stability of its operations lies in functional focus and constancy, detailed description of process, well defined job/work explanation and execution, bureaucratic controls, and well established lines of communication. On the other hand, information technologies provide a process focus, enable cross functional interaction and cooperation, flatter structures, and quicker communication. This difference has resulted in a friction that hampers alignment between technology and asset management goals and objectives. Information systems have not been institutionalised in Company A, and it is still struggling to adjust and adapt to them.

7.1.2   Road Infrastructure Asset Management – Case B

Case B presents a unique situation where a consortium is engaged in planning, design, operation, maintenance, and management of an infrastructure asset for a period of 40 years. This arrangement is driven by financial considerations rather than operational efficiency gains. This form of collaboration is not new in construction industry, since a usual construction project normally involves a number of different organisations that come together to execute the project. This collaboration ends when the project is completed. However, in terms of Case B the collaboration is required to work for an extended period of time. Even though the concept provides good grounds for drawing upon each organisation's core competencies and expertise, and risk sharing, it also introduces new problems and issues for information systems utilised for asset management.

The basic issue in Case B is the immaturity of the concept of collaboration in the consortium. As a result the infrastructure that enables collaboration is not mature. Although, a centralised technological platform, Optus inCITE, is being used for document exchange, storing, and management. However, it is not enough in meeting the demands of asset management. The available information is not integrated with administrative, financial, and operational information. As a result, management at Case B do not feel confident enough to use this information for decision support. It is also clear from the case that the organisations involved in the consortium are not willing to invest in integrating their information systems for two major reasons. Firstly, Optus inCITE is not an industry standard technology and the temporary nature of the project does not provide enough motivation for any organisation in the consortium to develop information integration architecture specific for this platform. Secondly, the question of information ownership has not been resolved. There is no shared responsibility of information at the consortium level, or at the individual organisation's level. As a result, the Optus inCITE repository contains information of varying quality, and with no process to audit it.

The consortium has a loose governance structure and its planning horizon focuses on a timeframe of around four years. As a result, information systems adoption also has a short term focus and is largely driven by the governance considerations of the consortium rather than asset lifecycle management. Even within the governance considerations, only the policy and procedural aspects are considered and context based social, organisational, and structural aspects that shape and institutionalise technology are ignored. This lack of commitment from top management and the inability of middle managers to plan for gaining advantages from technology has been less than inspiring for the functional staff to effectively utilise technologies like Optus inCITE. Consequently, the underlying mindset at Case B sees information systems as passive entities with little or no input into strategic asset management. It is, therefore, not surprising that technology implementation in Case B lacks strategic vision, and no consideration is given to the context within which it is to be implemented.

The consortium, due to its loose governance arrangements, has a reactive approach to technology implementation. Consortium organisations participating in the management of asset lifecycle decide on how the core asset management activities will be executed; however, enabling infrastructure like information systems is left to individual organisations to choose. Therefore, each participating organisation is more interested in using information systems for their own efficiency, and is not motivated to invest in technology for consortium wide cooperation and coordination. There is heavy reliance on inCITE platform for cooperation and coordination in the consortium, whereas, it is not capable and mature enough to provide the required level of functionality. At best, it could be used for document exchange and enabling some of the essential workflow. This situation has resulted in fragmentation of tasks, and organisations in the consortium are more concerned about accomplishing what has been assigned to them rather than what the consortium needs to accomplish. As a result, asset management in Case B is not driven by asset need/demand; it is actually driven by financial considerations, where each organisation is vying for their profit maximisation.

Since technology implementation and management is not a core responsibility of the consortium, training of employees on new technology is left to individual organisations. Consequently, there is varying degree of expertise found in the consortium, and the same variation is also reflected in process efficiency throughout the consortium. These inefficiencies pose significant risks to asset management due to the resulting inabilities to effectively use information systems to capture, share, manage, and analyse data. Although, individual organisations mine data, and preserve learnings and knowledge for their own reasons, yet there is no such initiative taken at the consortium level for asset lifecycle management. At the same time, this inability to utilise information systems properly is contributing to an inefficient integration of the asset value chain. At the moment all the emphasis is on asset design and construction, and there is no consideration given to later stages of asset lifecycle, such as asset operation and maintenance. Information systems are deployed and utilised to aid in design and construction, and there is no emphasis on capturing information that could be useful later in the asset lifecycle. It can be safely concluded that Case B will face significant issues in later stages of the asset lifecycle, such as asset capacity management, service provision under certain conditions, asset maintenance, and asset renewal and rehabilitation.

In summary, extemporized and unplanned nature of technology adoption is prevalent in case B. Technology implementation is opportunity driven and not motivated by the process needs. As a result, information lacks quality, is incomplete, and is of little strategic asset lifecycle management value. Managing project learnings is not on the strategic agenda of the consortium. Even in instances where individual organisations engage in preserving operational knowledge, they are not willing to share it with other organisations. The temporary nature of the consortium entails that the consortium sees no advantage in measuring the performance of their business processes or enabling technology and, therefore, does not do so.

7.1.3   Rail Infrastructure Asset Management – Case C

Case C resembles Case A in more than one way. Company C is also a semi government public sector organisation, has a hierarchical structure, bureaucratic culture, and relies heavily on centralised decision making. However, Company C is more technology savvy than Company A and has a relatively proactive attitude towards information systems utilisation. This utilisation could best be described as an emergent since Company C has experienced continuous change in the last 10 years. This change has primarily been necessitated by the Australian government's policy of deregulation and has also been assisted by stricter asset management regulation from the state government. As a result the organisation has to continuously adapt to the changes in the internal as well as external business environment.

Company C is relatively better placed than other public sector organisations with regard to management's commitment to information systems adoption. Its Chief Operating Officer sees IT as a business enabler and that is one of the major reasons that the organisation went through the recent technology mapping exercise. However, that exercise was carried out by external consultants and was restricted to high level information systems architecture. Furthermore the exercise was aimed at creating an overall framework from scratch while retaining the major technologies, rather than providing assessments into maturity of existing technologies enabling asset lifecycle management process. It also did not specify how the organisation might overcome issues relating to socio-technical system alignment, technological dependency of business processes, integration of information systems within the organisation, training issues, and other operational issues faced by functional staff (for example, data quality). The Asset Information Management Framework proposed by this technology mapping exercise (see section 6.4.2) has a high level of abstraction, discusses technology in terms of interfaces only, and disregards the soft benefits to be attainted from information systems implementation. In a nutshell, the evaluation effort has different meaning for different departments and stakeholders within the same organisation; whereas it should have been an exercise aimed at enabling mutual translation of strategic asset management objectives through the use of information systems by all organisational stakeholders.

Company C, being a bureaucratic setup, lays unnecessary details in policy and operating procedures definition. For example, the organisation has an assortment of different plans and policies regulating the use of technology, which include IT policy, IT governance plan, information systems plan, information policy, and information governance plan. These plans are developed by different stakeholders, with some developed by in functions/departments that do not come under the control of the CIO. It is, therefore, not surprising that these plans and policies are, for the most part, stand alone documents, are not connected to one another, and the trace between these documents and overall strategic business considerations cannot be easily established. This 'over governance' of information technology infrastructure has resulted in confusion. For example, multiple information systems record information about the same incident (see section 6.4.5). It has, in certain cases, also resulted in 'analysis paralysis' and has adversely affected decision support and process efficiency. It is for the same reasons that asset lifecycle processes are working with varying degree of efficiency, for example maintenance at Company C is much more established and efficient than other functions, such as asset design and renewal/refurbishment.

One of the major issues resulting from the 'technology over governance' in Company C is the lack of information ownership. There is no evidence of information stakeholders taking responsibility and ownership of its creation, access, modification, processing, storage, use, and reuse. The lack of information control, supervision, and management has compromised information integrity and credibility. This lack of responsibility in Company C has also affected information lifecycle management, for example, information producers or creators are only concerned with its creation and do not account for where, how, who, why, and for what purpose that information will be used. Since in Company C responsibilities are not assigned about data creation, processing, and maintenance, there is no accountability that would clearly define the rights, commitment, and responsibilities of organisational stakeholders with regard to information management. In the existing circumstances, efforts are aimed at recording information about what activities have taken place and are not focused towards creating information enabled and integrated asset lifecycle view.

Company C does not conform to any specific technology based enterprise architecture and as a result there are many ad-hoc information systems solutions in operation within the organisation. Technology is not mapped on to asset management processes' requirements, and in fact some of the processes in the organisation could be termed as over automated. This also brings to fore the issues relating to lack of information integration and interoperability. A by product of these issues is the fact that asset lifecycle information is not integrated and thus restricts realisation of an integrated view of lifecycle. As a result, Company C is unable to profile asset behaviour in financial and operational terms.

Middle management at Case C acknowledges the need of preserving asset lifecycle knowledge for better lifecycle decision support by sieving out important learnings on the effectiveness of existing asset lifecycle processes and management regimes. However, this has not been put into practice as the organisation lacks a consistent policy on technology implementation. Case C is using SAP as its core asset management technology, and hopes to interface all the other administrative and operational technologies around it. The choice of this technology was made by senior management. Their decision was heavily influenced by the reputation of technology and the assumption that it has worked well for similar businesses. The impact of this ill-planned technology implementation is still being felt, as each function within asset lifecycle management is struggling to adjust its processes to the chosen technology. The issues faced in adapting to SAP are more or less similar throughout the organisation. However, the organisation has never evaluated the performance of technology that supports asset management and thus these issues have never been recognised at a level that would warrant any corrective action. Even where concerns were raised by individual departments, the real issue remained hidden as the issues were attributed to departmental or functional inefficiencies. Each time managers found an issue, instead of fixing the cause in the existing technical infrastructure another piece of technology was introduced as a workaround to the issue, hence the number and variety of ad hoc solutions in the organisation.

In summary, even though the top management is committed to effective information systems utilisation, middle management lacks understanding of information systems effectiveness and their impact on continuous improvement of asset lifecycle management. There is a strong leaning towards establishing and evolving the use of operational technologies to manage asset lifecycle functions such as design, operation, and maintenance. Of course, this is due to the heavy representation of engineers in the middle to senior middle level management in the organisation. These managers are more concerned about their primary engineering job. Their emphasis is on accomplishing the tasks rather than learning from what tasks have been accomplished, how they have been accomplished, and how they ought to be accomplished. Collectively, this narrow approach impedes evolution and maturity of technology in the organisation. It is for the same reason that even though the organisation has invested in state of the art information systems, it has been unable to use them to their full potential and garner associated advantages.

7.1.4   Common Themes from the Case Studies

The following sections provide a cross case analysis and bring about the common themes on their competitiveness, how the case organisations are managing their assets, and how they are utilising information systems.

7.1.4.1 Asset Planning/Design

The assets investigated in this research generally have life span that is spread over a few decades. This means that the majority of these assets were designed in the first half of the twentieth century and, therefore, their design information is not available. Case studies also revealed that the design information of the recently developed assets is also far from being complete and is seldom used for other asset lifecycle processes such as asset maintenance, renewal, and refurbishments. In addition, analyses for asset supportability design are not taken seriously and as a result organisations are not able to predict the financial and non financial resource of their asset base spanning their lifecycle. The reasons behind this inability are the lack of requisite information, lacking quality of information to allow for such assessments, and the lack of the integration of existing information systems infrastructure to provide these assessments. Most of the assets investigated for this research outlast the professional lives of asset managers. Consequently, there is a clear distinction between the asset management philosophy of asset owners and asset managers. Asset managers have a short term focus of asset lifecycle, which impacts holistic decision support. This focus has set boundaries and limits, for example the information systems for design are solely for aiding the design job. Existing information systems do not aid in asset demand management, need assessment, or other parameters that could help in definition of asset design parameters. Asset design, therefore, is all about how engineers perceive it to be, since they do not have access to information on all asset lifecycle parameters, historical information, or asset operational profiles. Therefore, not only that new designs are completed as an isolated activity, but lifecycle supportability analysis and specification could not be completed either. However, the designers themselves do not share design information with other lifecycle functions.

7.1.4.2 Asset Operations

Asset operation along with maintenance is the most automated stage of the life-cycle of investigated assets. Case organisations generally had information systems in place for scheduling operations as well as monitoring asset operations; though this information systems utilisation was not free of issues in any of the organisations. Case A and C did not have complete inventory of their assets. The case studies revealed that asset configurations may be spread over an area of several hundred of kilometres and may have restricted access, such as being underground; whereas the available information systems have not matured enough to provide a complete assessment of the condition of entire asset configuration. However, the most crucial revelation of the case studies with regard to information systems for condition monitoring was the fact that the operational technologies used for condition monitoring were not integrated with administrative systems, like ERP systems. As a result, condition information could not be used for any other purpose (for example finding out the financial impact of asst failure and shut down) than just raising alarms and issuing failure notifications. Every organisation expressed its desire to be able to profile asset lifecycle behaviour; though none of the case organisations shared asset operations information with design and maintenance function for asset operation profiling and maintenance prediction. Due to this, there were no agreed performance indicators for operation and workload, and in instances where they were formulated by managers, they were seldom followed by functional staff.

7.1.4.3 Asset Maintenance

Maintenance has traditionally been the focus of asset managing organisations. However, development of an all inclusive maintenance system has yet to come to fruition. In the case organisations examined for this research, even though maintenance functions were the most technologically mature; yet the use of information systems in maintenance execution was far from being satisfactory. Particularly, recording maintenance information was not being taken seriously, and whatever information was collected lacked quality. Maintenance information was not integrated with financial information; thereby making it difficult to find out the complete (both evident and hidden) costs of asset failure, asset shutdown, as well as maintenance. In addition, this information was neither shared with other functions, nor was it reused to sieve out learnings that could be used in later maintenance cycles as well as at other asset installations of similar nature. Furthermore, maintenance execution was found to be heavily driven by the tacit knowledge of maintenance crew, and there was no effort made towards preserving that knowledge. Interestingly, two of the case organisations used SAP systems as their core asset lifecycle management system, yet none of them had fully implemented its plant maintenance module and relied on customised, stand alone systems for maintenance management.

7.1.4.4 Operational Efficiency

All the case studies revealed that investments in information systems were made to fix issues and problems in individual asset lifecycle processes, and did not have an entire asset lifecycle focus. Another common theme that emerged from these case studies was the clear lack of inter-organisational coordination. Due to this lack of collaboration there is hardly any motivation for sharing of knowledge and lessons learnt from successes and failures. The case organisations showed a general lack of an evaluative culture, and no recognition of social and cultural benefits that investments in information systems enable. Management was found wanting in need of enforcing an effective change management strategies, which would have created the realisation of need of technology within the organisation and would also have allowed for institutionalisation of technology. On the contrary, information systems utilisation in the case organisations is considered as an extra workload. As a result, instead of bringing the asset lifecycle value chain together, information systems have created islands of information. This issue has resulted in another challenge posed to all the organisations, which was the inability to manage lifecycle knowledge; whereas, all the organisations rely heavily on tacit knowledge of staff to execute critical lifecycle activities. Training is an area that was found to be quite weak in all the three cases. There was no structure for training in place and basically staffs were left to train themselves. Due to this, there are varying levels of use of information systems and equally fluctuating levels of quality of information that resides in these information systems.

7.1.4.5 Competitiveness

In all the organisations, technology implementation had a narrow focus and did not appear to be aimed at achieving any strategic advantages. There was a fragmented approach to information systems implementation, with scores of unsanctioned customised spreadsheets and databases. As a result information systems were not integrated at any level in any of the case organisations, due to which none of the organisations was able to profile asset operation in terms of efficiency and cost effectiveness. At the same time, there was no attempt made to use information systems and information contained in them to guide innovation, creativity, and competencies development in any area of lifecycle management. Information systems in all the organisations were focused on achieving internal efficiencies, and have little contribution in terms of making these organisations responsive to external forces.

In summary, all the organisations involved in the three cases faced issues with technology implementation and had no system in place to evaluate information systems based asset management processes, which could highlight underperforming areas and reveal the unmet information requirements of the asset management processes. Although, two of these organisations underwent evaluation by external evaluators; however those exercises did not focus on technology, were one-offs, and were not part of their regular business strategy execution cycle. At the same time, these evaluations were focused on quantitative tangible process outputs. They did not account for the means to achieving these outputs and their impact on other processes, and they also did not take into consideration the soft qualitative benefits of technology implementation. The findings from these case studies emphasise three areas for the success or failure of establishing or developing information systems based asset management evaluation methodologies. These three areas include the approach to information systems implementation, the roles of information systems in asset lifecycle management, and the role of context in institutionalising technology and shaping an evaluation methodology.

7.2 Aligning Information Systems with Strategic EAM

With this backdrop, this research now attempts to answer the first sub question, i.e. 'how do information systems facilitate alignment of strategic asset management considerations with overall business strategy and organisational design?'

The analysis of case studies revealed that none of these organisations conform to any specific information model or information systems architecture. As a result the approach to technology implementation is fragmented and narrow, which views technology as instruments for process automation rather than viewing it as a building block towards a holistic information systems based asset management regime. The manifestation of these issues can be found in the variances between the implementation approach and the continuously changing context within which technology is implemented. Consequently, these organisations have struggled to institutionalise and evolve the value profile of technology within their operating context.

Section 3.5.1 introduced an information systems based alignment framework (see figure 3-11 reproduced below). The framework takes a resource based view and highlights the technology implementation and its institutionalisation process through the mutual interaction of four domains. It accounts for the social shaping of technology and aligns the role of information systems within the organisation as well as with the organisational strategy. The framework seeks to develop alignment of gaols, intent, functions, and context through the utilisation of information systems. The framework is evaluation driven and emphasises evaluation of performance of asset management processes as well as the enabling information systems to keep them attuned to the strategic, social, and technical needs of asset management. The asset lifecycle management domain is, thus, strategically aligned with the organisational design domain in the sense that not only the asset lifecycle objectives are achieved but the way assets are managed contributes to the responsiveness of the organisation. In other words, asset management processes are coupled with organisational design, such that the organisational resources and infrastructure evolve and mature with asset management processes.

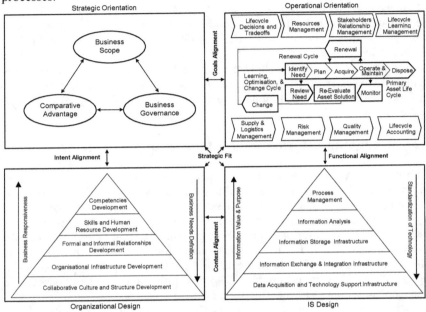

Figure 3-11: Information Systems Based EAM Alignment Framework

On the flip side, the asset management processes specify the level of maturity and the types of organisational infrastructure required to put in place a sound asset management regime. Furthermore, it requires asset lifecycle management processes to adapt to the changes in the internal as well as external business environment. This adaptation depends upon the speed and flexibility with which the organisation updates the use of the information systems and the meaning and use that it attaches to information systems utilisation in the context of asset management. In doing so it not only highlights the nature of technology to be implemented but also underscores the social and cultural process that facilitate technology institutionalisation by creating the shared meaning establishing the use of technology within the organisation.

The framework further addresses the information systems implementation issues identified in the case studies. It argues coupling of organisational design, information systems design, and asset operational orientation with strategic orientation of the business, and stresses derivation of asset lifecycle management processes from strategic business objectives. Technology is thus implemented in response to the information needs of asset management processes. Asset lifecycle management thus becomes information driven, since information requirements of asset management processes define the choice of technology. However, use and institutionalisation of technology is dependent on the context within which it is implemented. Consequently, it is essential to develop the organisational infrastructure such that it matures and evolves with technology and thus helps the organisation to adapt to changes in technology as well as changed brought about by technology (which may be necessitated due to a variety of internal as well as external reasons). The framework thus accounts for the hard factors such as staff training, job redesign, standardisation of practice, as well as soft factors such as culture and competencies development. Evaluation of information systems is embedded in the framework, which requires the organisation to assess suitability of technology to the asset management needs driven by business objectives. The organisation, thus, needs to continuously engage in a process of aligning information systems with the asset management processes that are drawn from the overall business strategy, and seek the maturity of the organisational design and infrastructure in the same process. In this case, each domain provides feedback to other domains and thrives on the same. This feedback allows for continuous improvement and enables the organisation to be responsive to changing business needs through evolving use of technology aimed at organisational maturity and competitiveness.

7.3   Factors Impacting Institutionalisation of IS Based EAM

This section addresses the second sub question, i.e. 'what factors impact institutionalisation of information systems based engineering asset management processes and their performance evaluation?'

The literature review and case studies reveal that information systems for asset management have dynamic multifaceted roles. These systems have to, among other uses, enable individual asset lifecycle business processes, provide for an integrated information enabled view of asset lifecycle to allow for informed decision support, facilitate organisational learning, and enhance competitiveness and responsiveness of the organisation. It is, however, difficult to account for all soft and hard factors that information systems enable, in a technology implementation or performance evaluation exercise. It is even more complicated for asset management, since it is aimed at assessing economic, operational, and strategic impacts of information systems.

The case studies reveal a range of factors that impact institutionalisation of information systems based engineering asset management processes and their performance evaluation, which could enable them to better understand the needs and performance of asset lifecycle management. These factors have human, technical, social, organisational, and procedural dimensions and impact development, adoption, and institutionalisation of information systems based asset management methodology in a variety of ways, these are,

a. Reactive rather than proactive approach to asset management, which is the major hurdle in effective long term planning for an effective asset management enabling infrastructure. Most of the technology adoptions are either in response to regulatory pressure or due to competitors adopting technology.

b. Technology implementation and planning is carried out independent of the context, as well as social, organisational, and technical maturity of the organisation. As a result, there is no fit between technology and business processes and the organisational infrastructure.

c. Deterministic approach to technology adoption, without introducing or facilitating changes in the structure and environment of the organisation.

d. Reliance on technology push rather than process information needs oriented technology pull.

e. Inability of the organisation to develop or conform to specific information model or information systems architecture.

f. Inability of the organisation to preserve asset lifecycle learnings and manage asset lifecycle knowledge.

g. Asset management is driven by financial considerations rather than asset need/demand.

h. Multiplicity of asset ownership within the same organisation resulting in multiple controls in critical areas.

i. Inability to integrate financial information with asset lifecycle information to find out the real cost of asset operation and shutdown.

j. Inability to develop an accountability based action oriented evaluative culture to measure the performance of information systems based asset management processes.

k. Lack of system and information integration and consequent failure to create an integrated view of asset lifecycle.

l. Lack of availability of complete historical information on asset lifecycle and heavy reliance on tacit knowledge of staff for executing core asset lifecycle actions.

m. Inability of the commercial off the shelf systems to provide the same functionality as is required by the organisation. It takes substantial time, effort, and resources to reengineer these systems as well as existing business processes to fit to these systems.

n. Strict hierarchical structure and communication lines with a silo approach to asset lifecycle management.

o. Tight coupling of technology with asset management processes, whereby increasing technology dependence for business processes execution.

p. Inability to verify efficiency, competence, and reliability assurance of asset lifecycle processes. This is further complicated by the inability to assign accountability, due to lack of control over customised ad-hoc solutions developed by different departments.

q. Inadequate inter departmental and intra organisational collaboration, which restricts quality of information and lack of confidence in IS.

r. Rigid application of information systems (such as the case with SAP in case A and C), which limits ability of the organisation to adapt to changing process requirements.

s. Inability to institutionalise technology in the field. Staff in the field (for example maintenance crew) do not take information systems seriously. Furthermore, the understanding among employees that their performance is judged on their primary job (such as operations, and maintenance) and not on how effectively they can use information systems.

t. Difficulties in implementing technology for condition monitoring in hazardous and complex situations.

u. Inadequate training and on job assistance available to utilise technology.

v. Maintenance execution and health information not exchanged with other asset lifecycle functions, which results in wastage of money and effort due to repetition of same issues among and within different asset installations.

w. Inability of the organisation to broadcast changes in policies and procedures throughout the organisation and to ensure that all the staff has understood the same.

x. Passive work methodologies impeding creativity, innovation, and motivation to use technology.

y. Inadequate change management strategies that contribute to employee lack of trust in technology.

z. Inability to introduce job redesigns and reward schemes to motivate employees towards effective information systems utilisation.

aa. Inability to assign roles and responsibilities for implementation and evaluation of asset lifecycle processes, so as to institutionalise performance evaluation.

bb. Political correctness driving external evaluations, and as a result assessments are biased in favour of management. Consequently, change strategies only focus on the operational level of the organisation.

cc. Inability to regularly update asset lifecycle plans and enabling infrastructure.

7.4   IS Based EAM Performance Evaluation

Having dealt with the issue of information systems implementation for asset life-cycle management, the next step is to resolve, 'how information systems based asset lifecycle management processes should be evaluated?' This section addresses this question in the light of the findings from the case studies and discussions in earlier chapters, and develops an information systems based asset management performance evaluation framework.

Information systems enable business processes at each stage of an asset life-cycle, and also help in shaping the organisational infrastructure and social environment. Section 2.2 suggested that information systems in asset management have three major roles; firstly, information systems are utilised in collection, storage, and analysis of information spanning asset lifecycle processes; secondly, information systems provide decision support capabilities through the analytical conclusions arrived at from analysis of data; and thirdly, information systems provide an integrated view of asset management through processing and communication of information and thereby lays the foundation for asset management functional integration. In doing so, information systems translate strategic asset management decisions through the planning and management consideration into operational actions. Information systems, thus, align technology with strategic asset management considerations and translate these considerations into action at the operational level. At the operational level information systems enable and support execution of core asset lifecycle processes. Execution of these processes generates information, which is used for analysis and evaluation on how well the information systems are enabling asset management processes at the operational level. These assessments provide decision support for corrective action or re-engineering of asset management plans and strategies. The case studies have revealed that information systems based asset management evaluation needs to measure whether these systems are translating strategic asset management considerations into action. At the same time, this evaluation should also highlight performance gaps, so that corrective action could be taken.

Conclusion can be derived from the case studies that it is essential that the learnings gained from information systems implementation and their evaluation be made available throughout the organisation, so that the organisation learns from its mistakes and grows internally as well as externally. However, the major problems involved in achieving this objective were identified as lack of information and system integration; lack of quality information; inability of the organisation to match technology implementation with business needs, and lack of learning based evaluative culture. This research developed an information enabled integrated asset management framework in section 3.1.2 (see figure 3-7). In the light of the observations from the case studies, this framework is updated to develop an information systems based asset management evaluation framework.

7.4.1   IS Based EAM Performance Evaluation Framework

The foremost finding from the case studies reveals that asset lifecycle management needs to be lifecycle learnings focused, such that each lifecycle stage draws from and contributes to it to create a learning based integrated view of asset lifecycle. Information enabled integrated asset lifecycle management, thus, is learning driven rather than IT/IS driven as described in the figure 3-7. This implies that information requirements of asset management should dictate planning, execution, and management of asset lifecycle rather than the technologies that enable asset lifecycle processes. The updated framework, thus, divides asset lifecycle into seven perspectives, i.e. competitiveness, design, operations, support, stakeholders, lifecycle efficiency, and learning perspective. It embeds aspects like data quality, integration, standardisation, and interoperability, and IT/OT integration into the frameworks through the connections between different perspectives. From top down, the framework assesses the usefulness and maturity of information systems in mapping the organisation's competitive priorities into asset design and reliability support infrastructure. The framework assesses the contribution and maturity of information systems through five further perspectives before informing the competitive priorities of the asset managing organisation. In doing so, the framework evaluates the role of information systems as strategic translators as well as strategic enablers of asset lifecycle management and enables generative learning. It implies that instead of just providing a gap analysis of the desired versus actual state of information systems maturity and contribution, it also assesses the information requirements at each perspective and thus enables continuous improvement through action oriented evaluation learnings. The following sections elaborate on these points.

7.4.1.1 Capacity and Demand Management

As discussed in section 2.2.1, asset need management is drawn from the strategic considerations of the organisation, which aims to align asset management strategy and plan with organisation's resources to best meet stakeholders' needs. These needs, however, are driven by the economic, social, and environmental constraints prevailing in the competitive environment of the business. In the core asset lifecycle, asset demand and capacity specifies the nature of assets as well as the types of support infrastructure required to ensure asset reliability throughout its lifecycle. These activities are information intensive and the integrity and value of their outputs is dependent upon the availability, speed, breadth and depth, and quality of information regarding competitive environment of the organisation. As has been evident from the case studies, asset managing organisations desire to have this broad information base to evaluate and assess asset demand. This demand specifies the type of assets to be commissioned or constructed, and the design of the support infrastructure for asset lifecycle management. The information systems employed at this stage must aid in translating the need of asset; the design of asset configuration; management of project to commission or construct the assets; and in performing trends and analysis to predict the operational behaviour, maintenance demands, and financial and non financial resources to ensure smooth asset operation over its lifecycle. At the same time, for existing assets the information systems must provide for decision support regarding improvements required in existing asset configuration to address customers' demands. The nature of this information is multifaceted and therefore requires scanning of the external business environment as well as taking into consideration the learnings gained over the years from managing assets employed by the organisation. For top management, the most important competitive measure of information systems employed at this level is how effective they are in managing business intelligence; so that the organisation uses the same to grow as well as to be responsive to the competitive pressures. The case studies revealed that the value profile that asset managers and designers attach to information systems at this level is the measure of how effective these systems are in aiding the design of the asset as well to predict the lifecycle maintenance demands and resources to ensure smooth asset operation over its lifecycle.

These options are arrived at after having considered a series of analysis that encompass the capability potential of the organisation and associated costs for ensuring reliability of the asset operation. Therefore, the effectiveness of information systems at this stage is in ensuring asset supportability and design reliability through the in-depth coverage of lifecycle supportability analysis, which provide a roadmap for the later stages of the asset lifecycle. For example, what type of maintenance regime should be put in place to keep the asset configuration in near original condition/specification, what third party arrangements should be in put in place to perform maintenance, specification of environmental constraints for asset operation, costs associated with supporting asset lifecycle, asset capacity, spares requirements, and training requirements.

Figure 7-1 encapsulates this value profile of information systems in translating strategic asset management considerations into asset and lifecycle support design.

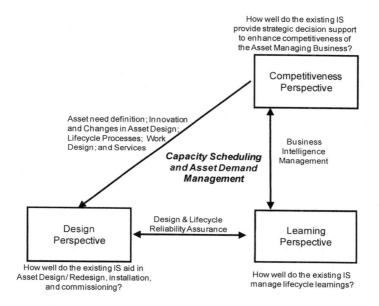

How well do the existing IS
provide strategic decision support
to enhance competitiveness of
the Asset Managing Business?

Competitiveness
Perspective

Asset need definition; Innovation
and Changes in Asset Design;
Lifecycle Processes; Work
Design; and Services

Business
Intelligence
Management

*Capacity Scheduling
and Asset Demand
Management*

Design
Perspective

Design & Lifecycle
Reliability Assurance

Learning
Perspective

How well do the existing IS aid in
Asset Design/ Redesign, installation,
and commissioning?

How well do the existing IS
manage lifecycle learnings?

Figure 7-1: Asset Capacity and Demand Management

In summary, at this level it is important to assess how information systems meet the demands of asset design and design for supportability of asset reliability, as well as their integration with other information systems in the organisation. An important consideration at this level, therefore, is the effectiveness of the organisation in utilising information systems to preserve asset design/redesign, capacity, scheduling, demand management/need assessment, and lifecycle prediction/profiling learnings and making them available to all other asset lifecycle management functions throughout the organisation.

7.4.1.2 Disturbance Management

Asset workload is defined according to its 'as designed' capabilities and capacity. However, during its operational life every asset generates some maintenance demands. Information systems that support asset operations specify asset operational schedule, asset workload assessment, condition monitoring, and disturbance management. The case studies revealed that, asset operators require the information systems at this level to refer to information regarding asset and lifecycle support design to develop operations plans and schedules, which means that operational information systems need to be integrated with systems that contain design information.

At the same time, in order to minimise asset operations disturbances, information systems utilised to aid asset operations must have access to lifecycle learnings that the organisation has accumulated over the life of its asset base. At this level, the role of operational technologies such as SCADA systems is critical. These operational technologies aid information systems in generation of consolidated health advisories by capturing and integrating condition information with asset health history, maintenance/treatment history, asset workload information, and design information. Such integration enables speedy generation of malfunction alarms and communication of failure condition information to maintenance function. As noted from the case studies, asset designers and maintainers also require disturbance information, therefore, it is important to assess if the existing information systems report back these errors to the asset design function so as to ensure asset design reliability. This information systems based arrangement is illustrated in Figure 7-2.

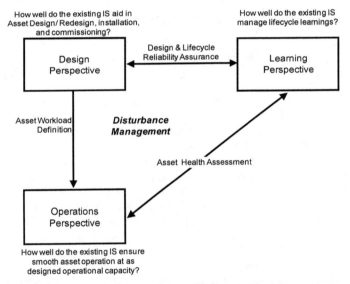

Figure 7-2: Health Assessments and Disturbance Management

In summary, at this level it is important to assess the value profile of information systems in ensuring smooth asset operation, condition monitoring and health assessments, and minimising and managing operational disturbances. At the same time, assessments on the integration of operations information system with design and maintenance systems provide a measure of availability and preservation of asset operation profiling and learnings to other functions in asset lifecycle management.

7.4.1.3 Asset Operation Risk Management

The notion of risk signifies the 'vulnerabilities' that asset operation is exposed to, due to the physical environment in which they operate or due to workload and operational conditions. The case studies revealed that asset managing organisations may have state of the art technologies to detect errors and failures, however, they generally fail to deliver the same level of efficiency when it comes to maintenance resources allocation, maintenance scheduling and workflow execution, and calculations of remnant lifecycle as well as the impact of asset shutdown on asset management as well as the overall business.

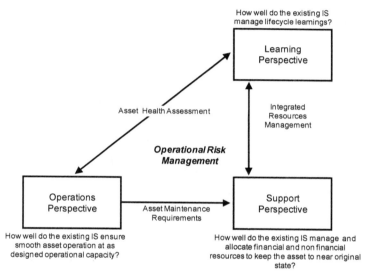

Figure 7-3: Operational Risk Management

The effectiveness of information systems at this level, therefore, needs to be assessed for their ability to provide complete assessment of the root cause analysis of failure condition, control and manage maintenance projects, ensure the availability of resources to carry out maintenance, and to execute maintenance/treatment workflow. However, the information and learnings from this level play a central role in other lifecycle management functions, as they have significant role in decisions regarding asset refurbishment, renewal, redesign, and retirement. Figure 7-3 illustrates the relationship between various information systems that enable maintenance and support activities as well as management of learnings generated from operational risk management. Therefore, at this it is important to assess the above mentioned value profile of information systems as well as their integration with operational technologies and other systems so as to provide maintenance information to other functions of the asset lifecycle management.

7.4.1.4 Asset Operation Quality Management

The aim of asset management is to keep the asset to near its original or 'as designed' state throughout its operational life. The case studies argued that once a disturbance has been identified, it becomes crucial to curtail its impact and to take appropriate follow up actions. These follow up actions not only involve direct actions taken on asset such as maintenance execution, but also involve sourcing of maintenance, rehabilitation, and renewal materials and expertise as well as enabling third party contractual agreements (Figure 7-4).

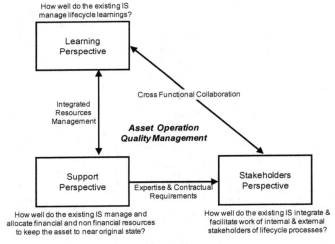

Figure 7-4: Asset Operation Quality Management

These third party arrangements are important, as more and more organisations are outsourcing maintenance to external organisations. However, these parties need to provide the asset managing organisations information regarding maintenance/treatment and the follow ups required. This information is critical for asset redesign/renewal/retirement. Environmental considerations are increasingly becoming important for asset managing organisations. It is, therefore, equally important to ensure that the asset operation conforms to the governmental and industrial regulations aimed at controlling the impact of disturbances on environment. Information systems at this stage have a versatile role, and therefore, it is important to assess how well the information systems enable collaboration and communication among various stakeholders to ensure effective cross functional relationships, quality maintenance and rehabilitation execution, timely availability of maintenance resources, as well as facilitating business relationships with external stakeholders and business partners.

7.4.1.5 Competencies Development and Management

It is evident from the case studies asset managing organisations generate enormous amount of explicit as well as tacit knowledge. The knowledge thus generated allows them to develop competencies in managing assets. These competencies when practised over time give rise to operational efficiency, which eventually contributes to the competitive advantage of the organisation. Nevertheless, information systems have the ability to capture and process knowledge and can also facilitate its sharing among organisational stakeholders. However, in order for this to happen it is important to find the task technology fit, since knowledge sharing requires suitable technology platform as much as it requires appropriate cultural, social, and personal values. At this level, as figure 7-5 suggests, information systems should be evaluated for how good they are in bringing together different stakeholders; so that they can share their knowledge and how good these systems are in aggregating, using and reusing, and managing organisational knowledge. This obviously requires assessing how good the existing information systems are in bringing together different functions of asset lifecycle management so as to enable an environment of knowledge sharing, such as enabling cross functional teams.

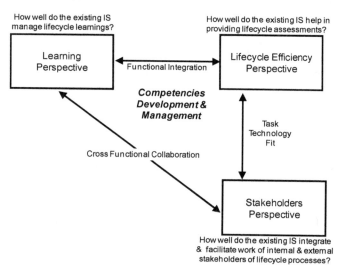

Figure 7-5: Competencies Development and Management

7.4.1.6 Organisational Responsiveness

Functional integration and a consolidated view of the asset lifecycle learnings facilitate the asset managing organisation in responding to the internal as well as external challenges. Information systems play an important role in materialising such responsiveness, by providing asset lifecycle assessments and profiles from operational, maintenance, financial, and non financial perspectives. These value assessments help the organisation in making strategic decisions, such as asset redesign, retirement, renewal, as well as cost benefits analysis of service provision and asset operation, and assessments of market demands. Nevertheless, the fundamental requirements in producing these value assessments are the availability of integrated and quality information that allows for an integrated view of asset lifecycle enabled by maintaining the asset lifecycle learnings (as figure 7-6 suggests). At this stage, information systems should be assessed for their contributions in providing business intelligence based decision support to strategic asset management; reporting in asset operational, lifecycle, and financial profiles; and customer/stakeholders demand assessment, and asset demand and need definition/redefinition.

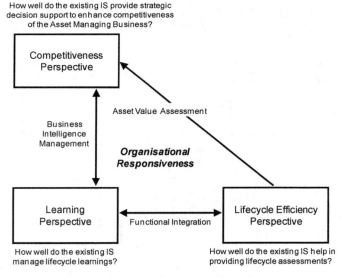

Figure 7-6: Organisational Responsiveness

Summing up the discussion, a consolidated framework for information systems based EAM evaluation is illustrated in figure 7-7.

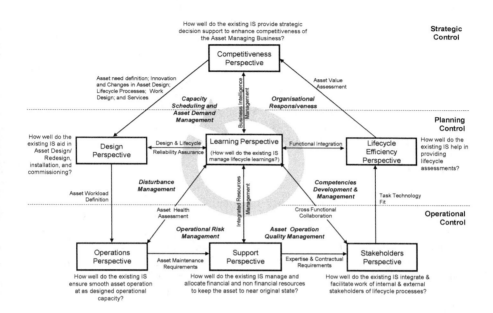

Figure 7-7: Information Systems Based EAM Evaluation Framework

This framework brings together all the perspectives discussed above. It is a learning centric framework that accounts for the core information systems based asset management processes as well as the allied organisational development areas that are influenced by information systems. It is context driven and accounts for the soft as well as the hard benefits gained from information systems utilisation in an asset lifecycle. The model defines and assesses the contribution of information systems through each perspective. At the design perspective, this model underscores the organisation's competitive priorities into asset design and reliability support infrastructure with the help of information systems. The model thus assesses the contribution and maturity of information systems through four further perspectives before informing the competitive priorities of the asset managing organisation. In so doing, the model translates asset management strategy into action through the use of technology. Furthermore, the assessment of the effectiveness of information systems in translation the strategic asset management concerns into action at each perspective provides the gap analysis between actual and desired state, which feed into the competitive perspective and thereby allow for the same information systems to be used as strategic translators as well as strategic enablers of asset lifecycle management. The whole exercise enables action oriented generative learning that facilitates continuous improvement of asset lifecycle management processes and enabling infrastructure. For example, this framework could be applied to assess any dimension of information systems utilised for asset management, such as strategic alignment, technical maturity, and information quality etc. This framework has generative learning is at the core, whereby it provides the results as set of actionable learnings so as to institute continuous improvement of asset lifecycle management processes and enabling technological infrastructure.

7.5   Summary

This chapter analysed the findings from empirical research, and the used the same to answer the research question set for this research. The findings from the case studies emphasise three areas for the success or failure of establishing or developing information systems based asset management regime and its evaluation. These three areas are the role of context in shaping the evaluation methodology; roles and responsibilities of the evaluators, and the approach to information systems implementation. With this backdrop, this chapter presented information systems alignment framework, which takes a resource based view and highlights the technology implementation and its institutionalisation process through the mutual interaction of four domains.

Having dealt with the issue of information systems implementation for asset lifecycle management, the chapter then attempted to resolve 'what factors impact institutionalisation of information systems based engineering asset management processes and their performance evaluation?' This chapter, thus, provides a range of factors identified from the analysis of empirical research. These factors have economic, operational, and strategic dimensions and illustrate what influences institutionalisation of information systems based asset lifecycle management. The chapter then resolves the question 'how information systems based asset lifecycle management processes should be evaluated?' It was argued that an information systems based asset management evaluation methodology needs to have a broad horizon and should account for assessment of the soft as well as hard benefits allowed by information systems. An information systems based asset management evaluation framework was thus developed that accounts for tangible and intangible aspects of information systems to asset lifecycle management.

The next chapter concludes this research. It highlights the practical and theoretical contributions made by this research. The chapter also provides insights into future research directions.

8 Conclusion

This research has attempted to investigate the issues related to information systems implementation for asset management and their evaluation. These issues have become increasingly important for asset managing organisations due to increased spending in IT, continuous managerial efforts towards better organisational resource utilisation, and the broad impact statement of IT in the organisation. These aspects, therefore, necessitate the study of information systems and information systems evaluation as part of the overall information systems governance and development for asset lifecycle management. However, it also needs to be acknowledged that there are numerous conceptual and operational difficulties that make investigation of these issues extremely complex and complicated.

8.1 Research Overview

8.1.1 Research Contributions
This research presents an analysis of the operational and conceptual issues posed to information systems based asset management evaluation. Therefore, this research contributes to both academic knowledge and industry practice. This research is of exploratory nature, as there is no pervious evidence of research into performance evaluation of information systems based asset management. However, there has been considerable evaluation activity in management and information systems literature, and through this research concepts from these disciplines were applied to asset management paradigm to generate new theories. Contributions of this research, therefore, can be summarised into three categories, i.e. the contribution to a new direction on technology evaluation research, development of information systems based asset management performance evaluation framework, information systems based asset management implementation, and the factors that may impact institutionalisation of information systems for asset management. Apart from these, this research also provides insights into asset management field in Australia through the three in-depth case studies.

8.1.2 Theoretical Contributions
The major theoretical contributions of this research are summarised below.

8.1.2.1 Literature Review
Since this is pioneer research in information systems based asset management, existing literature has been critically reviewed from a new perspective while

stressing the requirements of an interpretive evaluation methodology. Within the information systems and business performance evaluation paradigm, this research took a different stance than traditional positivistic and deterministic stance and approached the issue from a more broader horizon that includes social, technical, and organisational aspects of information systems based asset management evaluation. As a result, this research provides a sound foundation for further research in this area.

8.1.2.2 Theory Development

Information systems based asset management is context based and socially driven. Consequently an appropriate perspective is required to explain the complexities and intricacies of information systems based asset management. Traditionally, performance evaluation research has resulted in static evaluations and has seldom enabled actionable learning. The information systems based asset management performance evaluation framework described in section 7.4 provides a new cognitive perspective to the knowledge and significance of information systems based asset management. It allows action oriented evaluation that enables generative learning. This research has also developed a framework that aligns information systems with strategic asset management considerations in section 3.5.1. This framework accounts for the social shaping and institutionalisation of the information systems in the organisation.

The theoretical frameworks developed in this research are underpinned by application of previous research work in areas such as organisational behaviour, psychology, contextualisation, interpretivism in information systems, management, asset management and performance evaluation disciplines. In so doing, the high level concepts and conclusions were interpreted in low level building blocks of these frameworks. Consequently, this research has been able to clearly identify and provide in-depth explanations of the issues relating to the context, actors, content, processes, and the organisational and actor interactions in terms of asset management. The knowledge thus developed provides a rich understanding of the information systems implementation for asset management and evaluation issues in asset management organisational, social, and technical context. The base of knowledge developed through this research can provide a starting point for further research into information systems design and implementation approaches to information systems based asset management, integration, and governance.

8.1.2.3 Insights into Practical Asset Management

The three case studies provide in-depth understanding of the phenomenon of information systems implementation for asset management and its evaluation and allied areas. In addition to uncovering the intricacies of information systems based asset management, these case studies also highlight the technical maturity of the organisation with regard to the overall maturity of the industry that they operate in. In doing so, it puts forward the issues posed to information systems implementation and their performance evaluation and the impacts of failure of the case organisations to evaluate their information based asset management processes. The important learnings thus gained from these case studies underscore the need for an action oriented evaluation of information systems based asset management, where context and the strategic importance of evaluation is given due consideration. These lessons provide a roadmap for successful information systems implementation for asset management as well as emphasise actionable generative learning to facilitate continuous improvement.

8.1.3   Practical Contributions

This research has been carried out from an applied platform, i.e. the Cooperative Research Centre for Integrated Engineering Asset Management, which is a Commonwealth of Australia sponsored research initiative. Therefore, this research is focused on Australian asset management environment and has sound implications for asset managers, senior management, asset lifecycle decision makers, and information systems infrastructure managers and developers. This research provides information systems based asset management stakeholders with knowledge and frameworks to implement information systems and evaluate the performance of these systems in terms of translating the strategic asset lifecycle objectives into action as well as the role of these systems in enabling the same objectives. It provides information systems based asset management evaluators with action oriented generative learnings that allows for continuous improvement of information based asset management plans and strategies. An asset managing organisation would thus profit from the deliverables of this research through a rich understanding of the nature of information systems based evaluation as well as the roles and responsibilities associated with such evaluations.

The positive and realistic contributions of this research have been indicated by the initial feedback from the case study organisations. Particularly, Case A and C have incorporated the findings from the case studies, and have initiated a formal process of periodic information systems evaluation for asset management. Apart from this major initiative, these organisations are considering the recommendations of the research to update their current practices. However, considering the fact that these are public sector large organisations, it will take substantial time and effort to realise the true potential of these changes. Therefore, utilisations of the recommendations of this research are restricted to specific areas only. Although, Case B has not enforced any of the recommendations, however the consortium is reconsidering its choice of core technology. In addition, the findings of this research have been recognised as a helpful external evaluation.

The fundamental contribution that this research makes to business managers is the elicitation of practical aspects of context based information systems evaluation. The contextualisation of information systems evaluation makes it a social process and thus facilitates its institutionalisation in the evolution of the organisation through action oriented follow up, thereby aiding in realisation of a learning organisation. This allows for better information systems implementation and adoption strategies and clear definition of roles and responsibilities in institutionalisation of information systems in the workflow of the organisation. Such evaluations further enable a quality culture, allow for better utilisation of resources, facilitate a collective effort for process improvement, and aid in establishing of a creative and innovation based organisational culture. This research views information systems implementation as an emergent process rather than a static process that can be directed and controlled according to a set of predefined policies and assumptions. In doing so the implementation and adoption strategies continuously adapt to the changes in the internal as well as external environment of the organisation. Nevertheless, this also means that such conceptualisation is matched with equally broad vision of aspects such as organisational maturity, technology push and need pull dynamics within the organisation, and scope of managerial decisions. Only then the whole progression of information systems based process evaluation makes sense and creates value for the organisation.

Nevertheless, there are three key contributions that this research makes towards asset managers as well as asset management organisations. Firstly, while implementing information systems, traditionally asset managing organisations have only considered technical aspect and have ignored social, cultural, and organisational aspects. This research provides asset managing organisation with context based information driven information systems implementation and management framework, which accounts for the changing role of information systems as strategic enablers as well as strategic translators within the asset managing organisation. This framework proactively aligns asset management considerations with the business strategy and organizational infrastructure with the help of information systems. It advocates the fit between processes, information, and organisational infrastructure as the drivers of information systems implementation and institutionalisation, and signifies the role of information in increasing the responsiveness of the organisation to address the competitive challenges by aligning and re-aligning strategic business objectives with asset lifecycle goals.

Secondly, a generative learning based information systems based asset management evaluation framework, which allows asset managers to take stock of the existing information systems capabilities and process maturity. This evaluation then allows for the identification of right technological investment that satisfies the need pull as well as the allied areas that impact and are impacted by the introduction of technology. The generative learning based learnings provide for the continuous improvement of the asset management regime as well as the enabling information systems infrastructure.

Thirdly, this research provides a comprehensive set of factors that influence institutionalisation of information systems based engineering asset management processes and their performance evaluation. These factors have been discussed in detail based on their organisational, technical, and cultural dimensions. Thus the asset managers have a comprehensive set of recommendations available, which allows them to implement information systems based asset management as a routine management process within the organisation.

8.2   Research Limitations

8.2.1   Critique of the Adopted Research Paradigm

This research has adopted an interpretive research philosophy, and its suitability for this research has been discussed extensively in Chapter 5. However, every epistemological stance has some limitations and interpretive epistemology is no exception. Orlikowski and Baroudi (1991) summarise these limitations as, inability or limited ability to account for the extrinsic conditions that enable sense making; lack of appreciation of unplanned effects and impacts of activities; tendency to ignore the structural inconsistencies and disagreements within the organisation; and propensity to ignore explanation of evolution of change by methodically ignoring the factors that generate change.

This research is focused around a variety of organisational roles, characteristics, and settings. In particular, roles in an organisation themselves evolve with the passage of time and reflect the observations held by people assuming those roles at a particular moment in time. In addition, since humans perform those roles and their perception is bound to be influenced by factors such as, technical, organisational, and social context; work policies and procedures; and effects and side effects of historical change. It is possible that the researcher was ignorant of these changes and accepted the facts as they stood at the particular point in time.

Researchers (such as Galliers 1991a) argue that interpretive studies are influenced by their freeform and subjective disposition as well as from reliance on the ability of the researcher to identify partiality and pre-determined postulations. While investigating social phenomena such as information systems implementation for asset management and its evaluation, researcher cannot be disassociated from the social world and therefore the perception and observation of the researcher can be biased. It has been argued that (see for example Walsham and Waema 1994) even a conceptually rich research framework is unable to guide a researcher through the social process of research. In fact this issue has little to do with research framework, as it depends upon the social skills of the researcher. In such a situation, the independence of the researcher in terms of research process, context, and background are doubtful. Nevertheless, since this research is being undertaken within the Cooperative Research Centre for Integrated Engineering Asset Management, and the case organisations have a high stake in this initiative, it is possible that these factors might have affected the neutrality, fairness, and the critical character of case analysis. In addition the cultural background and limited experience of the researcher in information systems evaluation exercises may have influenced data analysis, and the conclusions arrived at from data analysis.

These issues and limitations were all known beforehand and suggestions and recommendations from previous research studies (see for example Farbey *et al.* 1995), were taken into account to take care of these issues. The research design was, therefore, cautiously and vigilantly designed (chapter 5). Some of the limitations of research design were taken care of during the research design stage. For example, triangulation of data sources was chosen to broaden the credibility of research results. However, a few issues were identified during field research, such as the inability of contextualisation in offering suitable justifications for the evolution of organisational culture and technological maturity. Other issues were the distinction between evaluation planners, evaluators, and users of evaluation results, which were often performed by same people. Apart from these, the various elements of conceptual framework (see section 7.4) overlap and are intertwined, whereas they were studied individually in this research. This isolation may affect the harmony of research findings. That is why instead of linking data to the framework, the information systems implementation for asset management and its evaluation framework were developed from interpretation of events.

8.2.2 Research Design Limitations
The case study methodology has also been subjected to criticism in terms of reliability of findings and their generalisability (Benbasat *et al.* 1987; Remenyi *et al.* 1998). Furthermore, the research findings may be criticised for lacking global validity due to limited number of case studies, i.e. three. Apart from this, the large size of the case study organisations may have restricted the generalisability of findings, since the study focused on specific departments and areas which may not be representative of the whole organisation. For example, in terms of Case B only the projects within the consortium were investigated, and in terms of C primarily the below track asset management was investigated. There were some issues encountered with the collection of empirical data. Permission was not always given to study the specific problem situation under investigation (for example, the evaluations carried out through Case A and C by employing external evaluators), but was granted to study the problem in a broad way. The use of different jargon, non familiarity with research literature and research process, and lack of knowledge of the need for research were some of the other issues

faced. This led to problems with achieving matching and attuned views on some issues.

8.3   Implications for Further Research

The research is of exploratory nature and sets the scene of information systems based asset management. It opens up a variety of avenues for further research. To begin with, further research in the same area could bring more insights by adopting a longitudinal case study method which could uncover the evolution of information systems implementation for asset management or their evaluation method, its institutionalisation within the organisation, its impacts and how organisations respond to and adapt to this change. It will be particularly useful to investigate how evaluation roles are developed. Another area could be the evaluation of maturity of information systems infrastructure for asset management and the level of its integration with the overall technological infrastructure of the organisation. While, yet another research direction could be the application of the findings of this research, particularly the information systems implementation and evaluation frameworks to other domains, such as health informatics.

This research presents an evaluation framework for the entire asset lifecycle. It will be useful to investigate function specific (such as maintenance) evaluation mechanisms with an expanded scope of technical, human, and cultural dimensions. This research has focused on core information technologies/systems. However, it will be interesting to apply the findings, conclusions, and outputs from this research to an environment where operational technologies are considered as a part of the organisation's overall information technology infrastructure.

References

AAA 2000, 'The Motorist's View in 2000', Australian Automobile Association, Report on October 2000 ANOP National Survey, Prepared by ANOP Research Services Pty Ltd, News South Wales, Australia

Abdel-Makoud, AB 2004, 'Manufacturing in the UK: contemporary characteristics and performance indicators', *Journal of Manufacturing Technology Management*, Vol 15, No. 2, pp.155-171.

Abdel-Malek, L, Das, SK, & Wolf, C 2000, 'Design and implementation of flexible manufacturing solutions in agile enterprises', *International Journal of Agile Management Systems*, Vol. 2, No. 3, pp. 187-195.

ABS 2005a, 'Business Use of Information Technology', Australian Bureau of Statistics, Catalogue No. 8019.0, Commonwealth of Australia, Canberra, ACT.

ABS 2005b '2005 Year Book Australia', Australian Bureau of Statistics, Number 87, Catalogue No. 1301.0, Commonwealth of Australia, Canberra, ACT.

ABS 2006, 'Australian National Accounts: Information and Communication Technology Satellite Account', Australian Bureau of Statistics, CAT No. 5259.0, 2002-03, Commonwealth of Australia, Canberra.

ABS 2007, 'Summary of IT Use and Innovation in Australia Business 2005-06', Australian Bureau of Statistics, Catalogue No. 8166.0, Commonwealth of Australia, Canberra, ACT.

Adam, A 2002, 'Exploring the gender question in critical information systems', *Journal of Information Technology*, Vol. 17, No. 2, 59.

Adams, R 1998, *Quality Social Work*, Macmillan, London.

Agarwal, R, & Sambamurthy, V 2002, 'Principles and Models for Organizing the IT Function', *MIS Quarterly Executive*, Vol. 1, No. 1.

Ahmed, AM 2002, 'Virtual integrated performance measurement', *International Journal of Quality & Reliability Management*, Vol. 19. No. 4, pp. 414-441.

Ahmed, JU 1996, 'Modern approaches to product reliability improvement', *International Journal of Quality & Reliability Management*, Vol. 13, No. 3, pp. 27-41.

AIG 2001, 'Industry in the Regions – 2001', Australian Industry Group, accessed online on November 26, 2006 at http://www.aigroup.asn.au/industryregions-overview.html.

Ajzen, I 1991, 'The theory of planned behavior', *Organizational Behavior and Human Decision Processes*, Vol. 50 No.2, pp.179-211.

Aladwani, AM 2002, 'An Integrated Performance Model of Information Systems Projects', *Journal of Management Information Systems*, Vol. 19, No. pp.185-210.

Alavi, M, & Leidner, DE 2001, 'Review: Knowledge Management and Knowledge Management Systems', *MIS Quarterly*, Vol. 25, No. 1, pp. 107-136.

Alexander, K 2003, 'A strategy for facilities management', *Facilities*, Vol. 21, No. 11/12, pp. 269 – 274.

Allen, JP 2000, 'Information systems as technological innovation', *Information Technology & People*, Vol. 13, No. 3, pp. 210-221.

Almgren, H 1999, 'Towards a framework for analyzing efficiency during start-up: An empirical investigation of a Swedish auto manufacturer', International Journal of Production Economics, Vol. 60-61, No. 1, pp. 79-86.

Alshawi, M, & Ingirige, B 2003, 'Web-enabled project management: an emerging paradigm in construction', *Automation in Construction*, Vol. 12, No. 4, pp.349-364.

Alstyne, MV, & Brynjolfsson, E 2005, 'Global village or cyber-balkans? modeling and measuring the integration of electronic communities', *Management Science*, Vol. 51, No. 6, pp.851.

Alter, S 2001, 'Are the fundamental concepts of information systems mostly about work systems?', *Communication of AIS*, Vol. 5, No. 11, pp.1-67.

Amadi-Echendu, JE 2004, 'The paradigm shift from maintenance to physical asset management', in Proceedings of 2004 IEEE International Engineering Management Conference, IEEE, Austin TX, Volume 3, pp. 1156-1160.

Amaratunga, D, Baldry, D, & Sarshar, M 2001, 'Process improvement through performance measurement: the balanced scorecard methodology', *Work Study*, Vol. 50, No. 5, pp. 179-189.

Anandarajan, A, & Wen, HJ 1999, 'Evaluation of information technology investment', *Management Decision*, Vol. 37, No. 4, pp. 329-339.

Anandarajan, M, & Arinze, B 1998, 'Matching client/server processing architectures with information processing requirements: A contingency study', *Information and Management*, Vol. 34, No. 5, pp. 265-274.

ANAO 2001, 'Management of the National Highways System Program', Australian National Audit Office, Department of Transport and Regional Services, Audit Report No.21 2000-2001 Performance Audit, Canberra, ACT, Australia.

Anderson, M, Banker, RD, & Hu, N 2002, *'Estimating the business value of investments in information technology*, in Proceedings of the Eighth Americas Conference on Information Systems, AMCIS 2002, Dallas, TX. pp. 1195-1197.

Anderson, T (ed) 1991, *The reflecting team: Dialogues and dialogues about the dialogues*, Norton, New York.

Andres, HP, & Zmud, RW 2001, 'A contingency approach to software project coordination', *Journal of Management Information Systems*, Vol.18, No. 3, 41-70.

Andresen, J 2001, 'A Framework for Selecting an IS evaluation Method - in the context of construction', Ph.D. Thesis, Technical University of Denmark.

Angell, I, & Smithson, S 1991, *Information Systems Management: Opportunities and Risks*. Macmillan, London.

Angell, LC, & Klassen, RD 1999, 'Integrating environmental issues into the mainstream: an agenda for research in operations management', *International Journal of Operations and Production Management*, Vol. 11, No.3, pp.63-76.

Angelo, IO, & Smithson, S 1991, *Information Systems Management*, Macmillan, London.

Angen, MJ 2000, 'Evaluating interpretive inquiry: Reviewing the validity debate and opening the dialogue', *Qualitative Health Research*, Vol. 10, No. 3, pp. 378-398.

Apostolopoulos, T, Doukidis G, & Pramataris, K 1997, *'A Techno-economic Approach for IT Investment Evaluation'*, in Proceedings of the 8th International Conference of the Information Resources Management Association, Vancouver, Canada.

Argyres, SN 1999, 'The impact of information technology on coordination: Evidence from the B-2 "stealth" bomber', *Organization Science*, Vol. 10, No. 2, pp. 162-180.

Argyris, C 1996, 'Skilled Incompetence', in *How Organizations Learn*, ed. K Starkey, International Thomson Business Press, London.

Argyris, C, & Schon, D 1978, *Organisational Learning: A Theory of Action Perspective*, Addison-Wesley, Reading, MA.

Argyris, C, & Schon, DA 1996, *Organizational Learning II: Theory, Method and Practice*, Addison-Wesley, Reading, MA.

Astrom, K, & Hagglund, T 2000, 'PID control', in Control system fundamentals, ed. W Levine, CRC Press, Boca Raton, FL.

Atkins, M, & Dawson, P 2001, 'The Virtual Organization: Emerging Forms of ICT-Based Work Arrangements', *Journal of General Management*, vol. 26, no 3, pp. 41-52.

Atkinson, AA 1998, 'Strategic Performance Measurement and Incentive Compensation', *European Management Journal*, Vol. 16, No. 5, pp. 552-561.

Atkinson, AA, Waterhouse, JH, & Wells, RB 1997, 'A stakeholders approach to strategic performance measurement', *Sloan Management Review*, Vol. 38, No. 3, pp. 25-37.

Atkinson, CJ 2000, 'The Soft Information Systems and Technologies Methodology (SIS-TeM): An actor network contingency approach to integrated development', *European Journal of Information Systems*, Vol. 9, No. 2, pp. 104-123.

Atkinson, H 2006, 'Strategy implementation: a role for the balanced scorecard?', *Management Decision*, Vol. 44, No. 10, pp. 1441-1460.

Austrade 2006, 'Railways overview', The Australian Trade Commission, accessed online on 12/12/2006 at http://www.austrade.gov.au/Railways-Overview/default.aspx.

Australian Infrastructure Report Card 2001, 'Australian Infrastructure Report Card', Engineers Australia, Barton, ACT.

Australian Infrastructure Report Card 2005, 'Australian Infrastructure Report Card', Engineers Australia, Barton, ACT.

Australian Railroad Group 2006, 'Australia's entire rail network', accessed online on December 11, 2006 at http://www.arg.net.au/images/ara.gif.

Bagchi, S, Kanungo, S, & Dasgupta, S 2003, 'Modelling use of enterprise resource planning systems: A path analytic study', *European Journal of Information Systems*, Vol. 12, No. 2, pp.142-158.

Bahli, B, & Rivard, S 2003, 'The information technology outsourcing risk: A transaction cost and agency theory-based perspective', *Journal of Information Technology*, Vol. 18, No. 3, pp. 211-221.

Bajaj, A, & Bradley, W 2005, '*An Exploration of the Role of Systems Analysis for Ex Ante Business Value Evaluations of Information Systems*', in Proceedings of the Eleventh Americas Conference on Information Systems, Omaha, NE, USA August 11th-14th.

Balch, WF 1994, 'An Integrated Approach to Property and Facilities Management', *Facilities*, Vol. 12, No. 1, pp. 17-22.

Ballantine, J, & Stray, SJ 1998, 'Financial appraisal and the IS/IT investment decision making process', *Journal of Information Technology*, vol. 13, no. 1, pp. 3-14.

Ballantine, J, Levy, M, Martin, A, Munro, I, & Powell, P 2000, 'An ethical perspective on information systems evaluation', *International Journal of Agile Management Systems*, Vol. 2, No. 3, pp. 233-241.

Ballantine, JA, & Stray, S 1999, 'Information systems and other capital investments: evaluation practices compared, *Logistics Information Management*, Vol.12, No. 1/2, pp.78-93.

Ballantine, JA, Galliers, RD, & Stray, SJ 1999, 'Information Systems/Technology Evaluation Practices: Evidence from UK Organisations', in *Beyond the IT Productivity*

Paradox, eds LP Willcocks, & S Lester, Chichester: John Wiley and Sons, pp. 123-150.

Balogun, O, Hawisa, H, & Tannock, J 2004, 'Knowledge management for manufacturing: the product and process database', *Journal of Manufacturing Technology Management*, Vol. 15, No. 7, pp. 575-584.

Bamber, CJ, Sharp, JM, & Hides, MT 1999, 'Factors Affecting Successful Implementation Of Total Productive Maintenance: A UK Manufacturing Case Study Perspective', *Journal of Quality in Maintenance Engineering*, 5(3), pp. 162-81.

Bamber, CJ, Sharp, JM, & Hides, MT 2000, 'Developing management systems towards integrated manufacturing: a case study perspective', Integrated Manufacturing Systems, Vol. 11, No. 7, pp. 454-461.

Banker, R, Robert, K, & Mahmood, MA 1993, 'A Comprehensive Bibliography of the Strategic and Economic Value of IT', in *Strategic Information Technology Management*, eds. in: R Banker, R Kauffman, & M A Mahmood, Idea Group Publishing, Harrisburg, SA, pp. 607-657.

Banville, C, & Landry, M 1989, 'Can the field of MIS be disciplined?', *Communications of the ACM*, Vol. 32, No. 1, pp. 48-60.

Barki, H, Rivard, S, & Talbot, J 2001, 'An Integrative Contingency Model of Software Project Risk Management, *Journal of Management Information Systems*, Vol. 17, No. 4, pp.37-69.

Barley, SR, & Kunda, G 2001, 'Bringing work back in', *Organization Science*, Vol. 12, No.1, pp.76-95.

Barrett, M, & Scott, S 2004, 'Electronic trading and the process of globalization in traditional futures exchanges: A temporal perspective', *European Journal of Information Systems*, Vol. 13, No. 1, pp.65-79.

Barry, B, & Crant, JM 2000, 'Dyadic communication relationships in organizations: An attribution/expectancy approach', *Organization Science*, Vol. 11, No. 6, pp. 648-664.

Bartels, A, Holmes, BJ, & Lo, H 2006, 'Global IT Spending and Investment Forecast, 2006 To 2007', *Forrester Research*, Cambridge, MA, USA.

Bartol, K, & Srivastava, A 2002, 'Encouraging knowledge sharing: the role of organizational reward systems', *Journal of Leadership and Organization Studies*, Vol. 9 No.1, pp.64-76.

Barton group 2005, 'Australian Water Industry Roadmap, Barton Group Environment Industry Development, Centre for Resources and Environmental Studies, Australian National University, Acton, ACT.

Basden, A 2002, 'The critical theory of Herman Dooyeweerd?', *Journal of Information Technology*, Vol. 17, No. 4, pp. 257-269.

Bausch, KC 2002, 'Roots and branches: a brief, picaresque, personal history of systems theory', *Systems Research and Behavioral Science*, Vol. 19, No. 5, pp. 417-428.

Beach, R, Muhlemann, AP, & Price, DHR 2000, 'Manufacturing operations and strategic flexibility: survey and cases', *International Journal of Operations & Production Management*, 20(1), pp.7-30.

Becerra-Fernandez, I, & Sabherwal, R 2001, 'Organization Knowledge Management: A Contingency Perspective', *Journal of Management Information Systems*, Vol. 18, No. 1, pp. 23-55.

Beckman, PA 2002, 'Concordance between task and interface rotational and translational control improves ground vehicle performance', *Human Factors*, Vol. 44, No. 4, pp. 644-653.

Bellgran, M 1998, 'Systematic Design of Assembly Systems', PhD Thesis, Department of Mechanical Engineering, Assembly Technology. University of Linkoping, Linkoping, Sweden.

Benbasat, I, Goldstein, DK & Mead, M 1987, 'The case research strategy in studies of information systems', *MIS Quarterly*, vol. 11, no. 3, pp.369-386.

Benjamin, R, & Scott Morton, M 1992, 'Reflections on Effective Application of Information Technology in Organizations From the Perspective of Management in the 90's', in Proceedings of the IFIP 12th World Computer Congress on Personal Computers and Intelligent Systems - Information Processing '92, Vol. 3, Amsterdam: North-Holland, pp. 131-142.

Bennett, D J, & Jenney, BW 1980, 'Reliability; its implication in production systems', *OMEGA The international Journal of management science*, 8(4), pp. 433-440.

Berry, WL, Bozarth, C, Hill, TJ, & Klompmaker, JE 1991, 'Factory focus: segmenting markets from an operations perspective', *Journal of Operations Management*, Vol. 10 No.3, pp.363-87.

Bertodo, RG 1989, 'On the deployment of automotive engineers', *Proceedings of the Institution of Mechanical Engineers (Part D: Journal of Automobile Engineering)*, Vol. 203 pp.15-23.

Bever, K 2000, 'Understanding Plant Asset Management Systems', *Maintenance Technology*, July/August, pp. 20-25

Beynon-Davies, P, Owens, I, & Williams, MD 2004, 'Information systems evaluation and the information systems development process', *The Journal of Enterprise Information Management*, Vol. 17, No. 4, pp. 276-282.

Bhatt, GD 2000, 'An empirical examination of the effects of information systems integration on business process improvement', *International Journal of Operations & Production Management*, Vol. 20, No. 11, pp. 1331-1359.

Bijker, WE, Hughes, TP, & Pinch, TJ (eds) 1987, *The social construction of technological systems: New directions in the sociology and history of technology*, MIT Press, Cambridge: MA.

Bijker, WE, & Law, J (eds.) 1992, *Shaping Technology/Building Society: Studies in Sociotechnical Change*, MIT Press, Cambridge, MA.

Bitichi, US 1994, 'Measuring your way to profit', *Management Decision*, 32(6), pp.16-24.

Bijker, WE 1995, *Of Bicycles, Bakelites, and Bulbs: Toward a Theory of Sociotechnical Change*, MIT Press, Cambridge, MA.

Bititci, US 2000, 'Dynamics of performance measurement systems', *International Journal of Operations & Production Management*, Vol. 20, No. 6, pp. 692-704.

Bititici, US, Carrie, AS, & McDevitt, L 1997, 'Integrated performance measurement systems: a development guide', *International Journal of Operations & Production Management*, Vol. 17, No. 5, pp. 522-534.

Bitichi, US, Nudurpati, SS, & Turner, TJ 2002, 'Web enabled performance measurement systems: management implications', *International Journal of Operations & Production Management*, Vol. 22, No.11, pp.1273-1287.

Bititci, US, Mendibil, K, Martinez, V, & Albores, P 2005, 'Measuring and managing performance in extended enterprises', International Journal of Operations & Production Management, Vol. 25, No. 4, pp. 333-353.

Bititci, US, Mendibil, K, Nudurupati, S, Garengo, P, & Turner, T 2006, 'Dynamics of performance measurement and organisational culture', International Journal of Operations & Production Management, Vol. 26 No. 12, pp. 1325-1350.

Bjork, BC 2002, 'The Impact of Electronic Document Management on Construction Information Management', in Proceedings of the International Council for Research and Innovation in Building and Construction, Council for Research and Innovation in Building and Construction Working group 78 conference 2002, 12 – 14 June, Aarhus School of Architecture. Aarhus, Denmark.

Bjorck, F 2004, 'Institutional Theory: A New Perspective for Research into IS/IT Security in Organizations', HICSS, p. 70186b, Proceedings of the 37th Annual Hawaii International Conference on System Sciences (HICSS'04) - Track 7.

Bjorn-Andersen, N 1988, 'Are 'Human Factors' Human?', The Computer Journal, Vol. 31, No. 5, October, pp.386-390.b

Blanchard, BS 1996, 'Life-cycle cost analysis: A technique for efficient asset management', in Proceedings of International Conference of Maintenance Societies 96 (ICOMS-96), Melbourne, pp. 31-38.

Blanchard, BS 1997, 'An enhanced approach for implementing total productive maintenance in the manufacturing environment', Journal of Quality in Maintenance Engineering, Vol 3, No. 2, pp. 69-80.

Blanchard, BS 1998, Logistics Engineering and Management, 5th edition, Prentice Hall, Upper Saddle River, NJ.

Blanchard, BS, & Fabrycky, WJ 1998, System Engineering and Analysis, 3rd edition, Prentice Hall, Upper Saddle River, New Jersey.

Bobbitt, LM, & Dabholkar, PA 2001, 'Integrating attitudinal theories to understand and predict use of technology-based self-service: The internet as an illustration', International Journal of Service Industry Management, Vol. 12, No. 5, pp. 423-450.

Bocij, P, Chaffey, D, Greasley, A & Hickie, S 2005, Business Information Systems: Technology, Development and Management for the E-Business, 3rd edn, Financial Times/Prentice Hall, New York.

Bock, GW, & Kim, YG 2002, 'Breaking the myths of rewards: an exploratory study of attitudes about knowledge sharing', Information Resource Management Journal, Vol. 15 No.2, pp.14-21.

Bock, GW, Zmud, RW, Kim, YG, & Lee, JN 2005, 'Behavioral intention formation in knowledge sharing: examining the roles of extrinsic motivators, social-psychological forces, and organizational climate', MIS Quarterly, Vol. 29 No.1, pp.87-111.

Boland, R J Jr. 1991, 'Information System Use as a Hermeneutic Process', in Information Systems Research: Contemporary Approaches and Emergent Traditions, H-E. Nissen, H. K. Klein, and R. A. Hirschheim (eds.), North-Holland, Amsterdam, 1991, pp. 439–464.

Boland, R J Jr., & Day, WF 1989, 'The Experience of Systems Design: A Hermeneutic of Organizational Action', Scandinavian Journal of Management, 5(2), pp. 87-104.

Boland, RJ Jr., & Hirschheim, RA eds. 1987, Critical Issues in Information Systems Research, John Wiley & Sons, Chichester.

Bolt, MA, Killough, LN, & Koh, HC 2001, 'Testing the interaction effects of task complexity in computer training using the social cognitive model', *Decision Sciences*, Vol. 32, No. 1, pp. 1-20.

Bose, R 2006, 'Understanding management data systems for enterprise performance management', *Industrial Management & Data Systems*, Vol. 106, No. 1, pp. 43-59.

Bostrom, RP, & Heinen, JS 1977, 'IS Problems and Failures: a socio-technical perspective', *MIS Quarterly*, (September), pp. 17-32.

Boudon, R 1986, *Theories of Social Change: A Critical Appraisal*, Polity Press, Cambridge.

Boulding, KE 1956, 'General Systems Theory: the skeleton of science', *Management Science*, Vol. 2, pp. 197-208.

Bowersox, DJ, Daugherty, PJ 1995, 'Logistics paradigms: the impact of information technology', *Journal of Business Logistics*, Vol. 16 No.1, pp.65-80.

Boyd, MJ 2001, '*A Recipe for Developing a Long Term Asset Replacement Plan*', in Proceedings of Distribution 2001Conference, Brisbane.

Boyer, KK, Ward, PT, & Leong, KG 1996, 'Approaches to the factory of the future: an empirical taxonomy', *Journal of Operation Management*, Vol. 14 No.4, pp.297-313.

Boyle, TA 2006, 'Towards best management practices for implementing manufacturing flexibility', *Journal of Manufacturing Technology Management*, 17(1), pp. 6-21.

Braam, GJM, & Nijssen, EJ 2004, 'Performance effects of using the balanced scorecard: A note on the Dutch experience', *Long Range Planning*, Vol. 37, No. 4, pp.335-349.

Braglia, M, Carmignani, G, Frosolini, M, & Grassi, A 2006, 'AHP-based evaluation of CMMS software', *Journal of Manufacturing Technology Management*, Vol. 17, No. 5, pp. 585-602.

Brennan, M 1992, 'Techniques for improving mail survey response rates', *Marketing Bulletin*, Vol. 3, pp. 24-37.

Brewer, J & Hunter, A 1989, *Multimethod research: A synthesis of styles*, Sage Publications, Newbury Park, CA.

Brignall, S, & Modell, S 2000, 'An institutional perspective on performance measurement and management in the New Public Sector', *Management Accounting Research*, Vol. 11, No. 3, pp. 281-306.

British Standard 1993, *Glossary of Terms Used in Terotechnology* BS 3811:1993, British Standards Institution Publishing Limited, London ISBN: 058022484 8.

Brown, CV, & Ross, JW 1996, 'The Information Systems Balancing Act: Building Partnerships and Infrastructure', *Information Technology and People,* 9(1), pp49–62.

Brown, MG Hitchcock, DE, & Willard, ML 1994, *Why TQM fails and what to do about it*, IRWIN Professional Publishing, Burr Ridge, New York.

Bruns Jr., WS, Kaplan, RS 1987, *Accounting and Management: Field Study Perspective*, Harvard Business School Press, Boston, MA.

Brynjolfsson, E 1993, 'The productivity paradox of information technology', *Communications of the ACM*, Vol. 36, No. 12, pp. 67-77.

Brynjolfsson, E 1994, 'Technology's true payoff', *Information week*, October, pp. 34-36.

Brynjolfsson, E, & Yang, S 1999, 'Intangible Benefits and Costs of Computer Investments: Evidence from the Financial Market (working paper), Massachusetts Institute of Technology, Sloan School, Cambridge, MA.

BTE 2000, 'Urban Congestion – the Implications for Greenhouse Gas Emissions', Information Sheet 16, Bureau of Transport Economics, accessed online on November 26, 2006 at http://www.btre.gov.au/docs/infosheets/is16.pdf.

Burgess, TF, McKee, D, & Kidd, C 2005, 'Configuration management in the aerospace industry: a review of industry practice', *International Journal of Operations & Production Management*, Vol. 25, No. 3, pp. 290-301.

Burke, K, Aytes, K, & Chidambaram, L 2001, 'Media effects on the development of cohesion and process satisfaction in computer-supported workgroups: An analysis of results from two longitudinal studies', *Information Technology and People*, Vol. 14, No. 2, pp. 122-141.

Burkhardt, ME 1994, 'Social interaction effects following a technological change: A longitudinal investigation', *Academy of Management Journal*, Vol. 37, pp.869-898.

Burrell, G, & Morgan, G 1979, *Sociological paradigms and organizational analysis*, Heinemann, London.

Busi, M, Bititci, US 2006, 'Collaborative performance management: present gaps and future research, *International Journal of Productivity and Performance Management*, Vol. 55, No. 1, pp. 7-25.

Callon, M 1986, 'The sociology of an actor-network: The case of the electric vehicle', in *Mapping the Dynamics of Science and Technology*, eds. M Callon, J Law, & A Rip, Macmillan Press, London.

Cameron, KS, & Quinn, RE 1988, 'Organizational paradox and transformation', In *Paradox and Transformation: Toward a Theory of Change in Organization and Management*, eds. RE Quinn & KS Cameron, Ballinger, Cambridge, MA.

Campbell, JD 1995, *Uptime: Strategies for Excellence in Maintenance Management*, Productivity Press, New York, NY.

Campbell, H 1996, 'A Social Interactinionist Perspective on Computer Implementation', *Journal of the American Planning Association*, Vol. 62, No. 1, pp. 99-107.

Cannel, E, & Nicholson, B 2005, 'Small firms and offshore software outsourcing: High transaction costs and their mitigation', *Journal of Global Information Management*, Vol. 13, No. 3, pp. 33-54.

Carson, D, Gilmore, A, Perry, C, & Gronhaug, K 2001, *Qualitative Marketing Research*, Sage Publications, London

Carter, S 1999, 'Anatomy of a qualitative PhD: part one getting started', *Management Research News*, Vol. 22, No.11, pp.9-22.

Castells, M 2001, *The Internet Galaxy: reflections on the Internet, Business and Society*, Blackwell. Oxford.

Chakravarthy, B 1997, 'A new strategy framework for coping with turbulence', *Sloan management review*, Vol. 38, No. 2, pp. 69-82.

Chan, CCA, & Scott-Ladd, B 2004, 'Organisational learning: Some considerations for human resource practitioners', *Asia Pacific Journal of Human Resources*, Vol. 42, No. 3, pp. 336-347.

Chan, FTS, Chan, MH, Lau, H, & Ip, RWL 2001, 'Investment appraisal techniques for advanced manufacturing technology (AMT): a literature review, *Integrated Manufacturing Systems*, Vol. 12, No. 1, pp. 35-47.

Chan, SC, & Lu, M 2004, 'Understanding internet banking adoption and use behaviour: a Hong Kong perspective', *Journal of Global Information Management*, Vol. 12, No. 3, pp. 21-44.

Checkland, P 1981, *Systems Thinking, Systems Practice*. John Wiley & Sons, Chichester.

Chen, ANK, & Edgington, TM 2005, 'Assessing value in organizational knowledge creation: Considerations for knowledge Workers', *MIS Quarterly*, Vol. 29, No. 2, pp.279-309.

Chen, JC, Chong, PP, & Chen, Y 2001, 'Decision criteria consolidation: A theoretical foundation of Pareto principle to Porter's competitive forces', *Journal of Organizational Computing & Electronic Commerce*, Vol. 11, No. 1, pp. 1-14.

Chen, Y, Chong, PP., & Chen, JC. 2000, 'Small business management: An IT-based approach', *Journal of Computer Information Systems*, Vol. 41, No. 2, pp. 40-47.

Chiang, L, Russell, E, & Braatz, R 2001, Fault detection and diagnosis in industrial systems, Springer-Verlag, London.

Chin, WW, Marcolin, BL, & Newsted, PR 2003, 'A partial least squares latent variable modelling approach for measuring interaction effects: results from a Monte Carlo simulation study and an electronic-mail emotion/adoption study', *Information Systems Research*, Vol. 14, No. 2, pp. 189-217.

Chung, WY, Fisher, CW, & Wang, RY 2005, 'Redefining the scope and focus of information quality work: a general systems theory perspective', in *Advances in Management Information Systems*, eds. RY Wang, WM Pierce, SE Madnick, & CW Fisher, ME Sharpe Inc., Armonk, NY.

Churchman, CW 1994, 'Management Science: Science of Managing and Managing of Science', *Interfaces*, Vol. 24, Issue 4, pp. 99-110.

CIA 2006, The World Fact book - Australia, Central Intelligence Agency, accessed online on October 23, 2006, at https://www.cia.gov/cia/publications/factbook/print/as.html.

Cibora, CU, & Lanzara, G 1994, 'Formative Contexts and IT: understanding the dynamics of innovation in organizations', *Accounting, Management and Information Technology*, Vol. 4, No. 2, pp. 61-86.

Ciborra, C 1996, 'Improvisation and Information Technology in Organizations', in proceedings of the ICIS, Cleveland, USA, December.

Clarke, P 1995, 'Non-financial measures of performance in management', *Accountancy Ireland*, Vol. 27, No. 2, pp. 22-24.

Clemons, EK, & Hitt, LM 2004. Poaching and the misappropriation of information: Transaction risks of information exchange. *Journal of Management Information Systems*, Vol. 21, No. 2, pp. 87-107.

Cohen, W, & Levinthal D 1990, 'Absorptive capacity: a new perspective on learning and innovation', *Administrative Science Quarterly*, Vol. 35, No. 1, pp 128-152.

Compeau, D, Higgins, CA, & Huff, S 1999, 'Social cognitive theory and individual reactions to computing technology: A longitudinal study', *MIS Quarterly*, Vol. 23, No. 2, pp. 145-159.

Cooper, RB, & Wolfe, RA 2005, 'Information processing model of information technology adaptation: An intra-organizational diffusion perspective', *Database for Advances in Information Systems*, Vol. 36, No. 1, pp. 30-48.

Cornford, T, & Smithson, S 1996, *Project Research in Information Systems: A Student's Guide*. Macmillan, London.

Costa, AS 1996, 'Economic evaluation of Information Systems: A Portuguese case study', in Proceedings of 4th European Conference on Information Systems, Ficha Tecnica, Lisbon, Portugal.

Cronk, MC, & Fitzgerald, EP 1999, 'Understanding IS business value: derivation of dimensions', *Logistics Information Systems*, Vol. 12, No. 1/2, pp. 40-49.

Cross, KF, Lynch, RL 1992, 'For good measure', *CMA Magazine*, 66(3), pp.20-23.

Curado, C 2006, 'Organisational learning and organisational design', *The Learning Organisation*, Vol.13, No. 1, pp.25-48.

Dabhilakar, M, & Bengtsson, L 2002, 'The Role of Balanced Scorecard in Manufacturing: a Tool for Strategically Aligned Work on Continuous Improvements in Production Teams', in *Performance Measurement and Management Control: a Compendium of Research*, eds. in MJ Epstein, & JF Manzoni, Elsevier Science Ltd., Oxford, UK.

Daft, RL, & Weick, KE 1984, 'Towards a Model of Organizations as Interpretation Systems', *Academy of Management Review*, Vol. 9, No. 2, pp.284-295.

Dahlbom, B, & Mathiassen, L 1993, *Computers in Context The Philosophy and Practice of Systems Design*, 2000 edition, Blackwell, Oxford.

Dangayach, GS, & Deshmukh, SG 2001, 'Manufacturing strategy: literature review and some issues', *International Journal of Operations and Production Management*, Vol. 21, No.7, pp.884-932.

Dangayach, GS, & Deshmukh, SG, 2005, 'Advanced manufacturing technology implementation - Evidence from Indian small and medium enterprises (SMEs)', *Journal of Manufacturing Technology Management*, Vol. 16, No. 5, pp. 483-496.

Daniel, EM & Wilson, HN 2003, 'The role of dynamic capabilities in e-business transformation', *European Journal of Information Systems*, 12(4), pp.282-296.

Das, SK, & Patel, P 2002, 'An audit tool for determining flexibility requirements in a manufacturing facility', *Integrated Manufacturing Systems*, 13(4), pp.264-274.

Davenport, TH 1998, 'Putting the Enterprise into the Enterprise System', *Harvard Business Review*, July-August, pp. 121-131.

Davern, M, & Kauffman, R 2000, 'Discovering Potential and Realizing Value from Information Technology Investments', *Journal of Management Information Systems*, vol. 16, no. 4, pp. 121-143.

Davies, A 1990, *Management guide to condition monitoring in manufacturing*, The Institution of Production Engineers, Holbrook & Son, London.

Davies, A, & Kochhar, A 2002, 'Manufacturing best practices and performance studies: a critique', *International Journal of Operations and Production Management*, Vol. 22, No.3, pp.289-306.

Davies, L, & Mitchel, G 1994, 'The Dual Nature of the Impact of IT on Organizational Transformations', in Proceedings of the IFIP WG8.2 Working Conference on Information Technology and New Emergent Forms of Organization, Ann Arbor, Michigan, USA, 11-13 August, 1994, eds. R Baskerville S Smithson O Ngwenyama & J DeGross, North-Holland, New York.

Davis, S, Albright, T 2004, 'An investigation of the effect of balanced scorecard implementation in financial performance', *Management Accounting Research*, Vol. 15, No.2, pp.135-153.

DCITA 2005, 'Achieving Value from ICT: Key Management strategies' ICT research study', Department of Communications Information Technology and the Arts,

Commonwealth of Australia Australian Government, April 2005, Opticon and Australian National University; Australian ICT Research Study, Canberra

de Geus, A 1996, *The Living Company*, Harvard Business School Press, Boston, MA.

De Sanctis, G, & Poole, MS 1994, 'Capturing the Complexity of Advanced Technology Use', *Organization Science*, Vol. 5, No. 2, pp. 121-147.

De Toni, A, & Tonchia, S 2001, 'Performance measurement systems - Models, characteristics and measures, *International Journal of Operations & Production Management*, Vol. 21, No. 1/2, pp. 46-70.

Deetz, S 1996, 'Describing Differences in Approaches to Organization Science: Rethinking Burrell and Morgan and Their Legacy', *Organization Science*, Vol.7, No.2, pp. 191-207.

DeLone, W H, & McLean, ER 1992, 'Information systems success: The quest for the dependent variable', *Information Systems Research*, Vol. 3, No. 1, 60-95.

DeLone, WH, & McLean, ER 2003, 'The DeLone and McLean model of information systems success: a ten-year update', *Journal of Management Information Systems*, Vol. 19, No.4, pp.9-30.

Dennis, AR, & Garfield, MJ 2003, 'The adoption and use of GSS in project teams: Toward more participative processes and outcomes', *MIS Quarterly*, 27(2), pp. 289.

Dennis, AR, Wixom, BH, & Vandenberg, RJ 2001, 'Understanding fit and appropriation effects in group support systems via meta-analysis", *MIS Quarterly*, Vol. 25, No. 2, pp.167-193.

Denzin, K 1978, *The Research Act*, McGraw-Hill. New York.

Denzin, N 1970, 'Strategies of multiple triangulation', in *The research act in sociology: a theoretical introduction to sociological method*, ed. N Denzin, McGraw-Hill, New York, pp. 297-313.

Denzin, NK & Lincoln, YS 1994b, *Handbook of QualitativeResearch*. Sage, London.

Denzin, NK, & Lincoln, YS 1994a, 'Introduction: Entering the field of qualitative research', in *Handbook of qualitative research*, eds. NK Denzin & YS Lincoln, Sage, Thousand Oaks, CA, pp. 1-17.

Dettmer, HW 1997, *Goldratt's Theory of Constraints: A system approach to continuous improvement*, McGraw-Hill, New York, NY.

Devaraj, S, & Kohli, R 2002, *Measuring the Business Value of Information Technology Investments*, 1st edn, Financial Times Prentice Hall, New York, NY.

Diaz, MS, Gil, MJA, & Machuca, JAD 2005, 'Performance measurement systems, competitive priorities, and advanced manufacturing technology: Some evidence from the aeronautical sector', *International Journal of Operations & Production Management*, Vol. 25 No. 8, pp. 781-799.

Dickinson, T, Saunders, I, & Shaw, D 1998, 'What to measure about organization performance', *Quality Magazine*, Vol. 7, pp. 71–78.

DiMaggio, PJ, and Powell, WW 1983, 'The Iron Cage Revisited: institutional isomorphism and collective rationality in organizational fields', American sociological review, 48 (2), 147-160.

Dodgson, M 1993, 'Organizational learning: A review of some literatures', *Organization Studies*, Vol. 14, No. 3, pp. 375-394.

Doherty, N, & King, M 2004, 'The Treatment of Organisational Issues in Systems Development Projects: The Implications for the Evaluation of Information Technology Investments', *Electronic Journal of Information Systems Evaluation*, Vol. 4, No. 1, accessed online on July 5, 2006, available at http://www.ejise.com/volume-4/volume4-issue1/issue1-art6.htm.

Donnellan, E 1995, 'Changing perspectives on research methodology in marketing', *Irish Marketing Review*, Vol. 8, pp.81-90.

Duberley, J, Johnson, P, Cassell, C, & Close, P 2000, 'Manufacturing change: the role of performance evaluation and control systems', *International Journal of Operations & Production Management*, Vol. 20 No. 4, pp. 427-441.

Duffuaa, SO, Ben-Daya, M, Al-Sultan, K, & Andijani, A 2001, 'A Generic Conceptual Simulation Model For Maintenance Systems', *Journal of Quality in Maintenance Engineering*, vol. 7, no. 3, pp. 207-219.

Duffuaa, SO, Cambel, JD, & Raouf, A, 1999, *Planning and Control of Maintenance Systems: Modelling and Analysis*, John Wiley & Sons, New York.

Dumond, EJ 1994, 'Making Best Use of Performance-Measures and Information', *International Journal of Operations & Production Management*, 14(9), pp. 16-31.

Dunn, C, & Grabski, S 2001, 'An investigation of localization as an element of cognitive fit in accounting model representations', *Decision Sciences*, 32(1), pp. 55-94.

Dwight, R 1999, 'Frameworks for measuring the performance of the maintenance system in a capital intensive organisation', PhD thesis, Department of Mechanical Engineering, University of Wollongong, Wollongong.

Earl, MJ 1989, *Management Strategies for Information Technology*, Prentice-Hall, Hemel Hempstead, England.

Earl, MJ 1993, 'Experiences in information systems strategic planning', *MIS Quarterly*, Vol. 17, No.1, pp.1-24.

Earl, M 1998, 'Integrating IS and the Organization: a framework of organizational fit', in *Information Management: the organizational dimension*, ed. MJ Earl, Oxford University Press, Oxford.

Edvardsson, B, Thomasson, B, & Ovretveit, J 1994, *Quality of Service*, McGraw-Hill, London.

Eerens, EWJ 2003, *Business Driven Asset Management for Industrial and Infrastructure Assets*, 1st edition, Le Clochard Publishers, Victoria, Australia.

Ehrhart, T 2002, 'All Wound Up: Avoiding Broken Promises in Technology Projects' *Risk Management*, vol. 49, no. 4, pp.12-16.

Eisenhardt, KM 1989, 'Building Theories from Case Study Research', *Academy of Management Review*. Vol.14, No.4, pp. 532-550.

El Hayek M, Voorthuysen, EV, & Kelly, DW 2005, 'Optimizing life cycle cost of complex machinery with rotable modules using simulation', *Journal of Quality in Maintenance Engineering*, Vol. 11, No. 4, pp-333-347.

El-Haram, M 1995, 'Integration approach to condition-based reliability assessment and maintenance planning', Ph D. Thesis, University of Exeter, Exeter, UK.

Ericsson, J. (1997). *Disturbance analysis of Manufacturing systems - A important tool for lean production*, PhD Thesis, Department of Production and Materials Engineering, Lund University. Lund, Sweden.

Ettlie, JE 1988, *Taking Charge of Manufacturing*, Jossey-Bass, San Francisco, CA.

Fabrycky, WJ, & Blanchard, BS 1991, *Life-cycle Cost and Economic Analysis*, Prentice-Hall, Upper Saddle River, NJ.

Farbey, B, Land, F & Targett, D 1993, *How to Assess your IT Investment. A study of Methods and Practice*, Butterworth Heinemann, Oxford.

Farbey, B, Land, F & Targett, D 1994, 'Matching an IT project with an appropriate method of evaluation: a research note on 'Evaluating investments in IT'. *Journal of Information Technology*, vol. 9, pp.239-243.

Farbey, B, Land, F, & Targett, D 1995, 'Evaluating business information systems: Reflections on an empirical study', *Information Systems Journal*, 5, pp. 235-252.

Farbey, B, Land, F & Targett, D 1999, 'Moving IS evaluation forward: learning themes and research issues', *Journal of Strategic Information Systems*, vol. 8, pp.189-207.

Fasheng, Q, & Teck, YK 2000, 'IS/IT Project Investment Decision Making', in Proceedings of the IEEE International Conference on Management of Innovation and Technology, ICMIT, Vol. 2, pp. 502-507.

Fattahi, R, & Afshar, E 2006, 'Added value of information and information systems: a conceptual approach', *Library Review*, Vol. 55, No. 2, pp. 132-147.

Feeley, TH, & Barnett, GA 1996, 'Predicting employee turnover from communication networks', *Human Communication Research*, Vol. 23, No.1, pp. 370-387.

Feurer, R, & Chaharbaghi, K 1995, 'Performance Measurement in Strategic Change', *Benchmarking for Quality, Management & Technology*, Vol. 2, No. 2, pp. 64-83.

Fiol, CM, & Lyles, MA 1985, 'Organizational learning', *Academy of Management Review*, Vol. 10, No. 4, pp. 803-813.

Foster, ST Jr., Sampson,SE, & Dunn, SC 2000, 'The impact of customer contact on environmental initiatives for service firms, *International Journal of Operations and Production Management*, Vol. 20, No. 2, pp. 187-203.

Fredendall, LD, Patterson, JW, Kennedy, WJ, & Griffin, T 1997, 'Maintenance: modeling its strategic impact', *Journal of managerial issues*, Vol. 9, No. 4, pp. 440-448.

Frisk, E, & Planten, A 2004, 'Evaluating IT: Learning from the past to design the future', in Proceedings of IRIS27, August 14-17, Falkenberg, Sweden.

Frisk, E 2007, 'Categorization and overview of IT perspectives – A literature review', in *Proceedings of the European Conference on Information management and evaluation*, University of Montpellier 1, Montpellier, France.

Gabbar, HA, Yamashita, H, Suzuki, K, & Shimada, Y 2003, 'Computer-aided RCM-based plant maintenance management system, *Robotics and Computer Integrated Manufacturing*, Vol. 19, No. 5, pp. 449–458.

Gable, G 1994, 'Integrating case study and survey research methods: an example in information systems', *European Journal of Information Systems*, 3(2), pp 112-126.

Galliers, RD 1991a, 'Strategic Information Systems: myths, realities and guidelines for successful implementation', *European Journal of Information Systems*, Vol. 1, No. 1, pp. 55-64.

Galliers, RD 1991b, 'Choosing appropriate information systems research approaches: A revised taxonomy', in *Information Systems Research: Contemporary Approaches and Emergent Traditions*, eds. HE Nissen HK Klein & R. Hirschheim, Elsevier Science Publishers, North-Holland, pp. 327 – 345.

Galliers, RD, & Land, FF 1987, 'Choosing an appropriate information systems research methodology', *Communications of the ACM*, Vol. 30, No. 11, pp. 900-902.

Galloway, I 1996, 'Design for support and support the design: integrated logistic support - the business case', *Logistics Information Management*, Vol. 9, No. 1, pp. 24-31.

GAO 2004, 'Water Infrastructure: Comprehensive Asset Management Has Potential to Help Utilities Better Identify Needs and Plan Future Investments', GAO-04-461, United States General Accounting Office, US government, Washington DC.

Garengo, P, Biazzo, S, & Bititci, US 2005, 'Performance measurement systems in SMEs: A review for a research agenda', *International Journal of Management Reviews*, Vol.7, No.1, pp.25-47.

Garg, A, & Deshmukh, SG 2006, 'Maintenance management: literature review and directions', *Journal of Quality in Maintenance Engineering*, 12(3), pp. 205-238.

Garicano, L, & Kaplan, SN 2001, 'The effects of business-to-business E-commerce on transaction costs', *Journal of Industrial Economics*, Vol. 49, No. 4, pp. 463-485.

Garrity, EJ 2002, 'Synthesizing user centred and designer centred is development approaches using general systems theory', *Information Systems Frontiers*, Vol. 3, No. 1. pp. 107-121.

Garvin, D 1993, 'Building a learning organization', *Harvard Business Review*, Vol. 71, No. 4, pp. 78-92.

Gattiker, TF, & Goodhue, DL 2005, 'What happens after ERP implementation: Understanding the impact of inter-dependence and differentiation on plant-level outcomes', *MIS Quarterly*, Vol. 29, No. 3, pp.559-585.

Gebauer, J, & Shaw, MJ 2004, 'Success factors and impacts of mobile business applications: results from a mobile e-procurement study', *International Journal of Electronic Commerce*, Vol. 8, No.3, pp.19-41.

Gerwin, D, & Kolodny, H 1992, *Management of Advanced Manufacturing Technology: Strategy, Organisation, and Innovation*, John Wiley & Sons, New York, NY.

Ghalayini, AM, & Noble, JS 1996, 'The Changing Basis of Performance Measurement', *International Journal of Operations & Production Management*, 16(8), pp. 63-80.

Ghalayini, AM, Noble, JS, & Crowe, TJ 1997, 'An integrated dynamic performance measurement system for improving manufacturing competitiveness', *International Journal of Production Economics*, Vol. 48, pp.207–225.

Gibb, S 2002, *Learning and Development: Process, Practices and Perspectives at Work*, Palgrave, London.

Giddens, A 1984, *The Constitution of Society: Outline of the Theory of Structure*, University of California Press, Berkeley, CA.

Ginberg, MJ 1980, 'An organizational contingencies view of accounting and information systems implementation', *Accounting, Organizations & Society*, 5(4), pp. 369-382.

Gindy, NNZ, Cerit, B, & Hodgson, A 2006, 'Technology roadmapping for the next generation manufacturing enterprise, *Journal of Manufacturing Technology Management*, Vol. 17, No. 4, pp. 404-416.

Gits, CM 1994, 'Structuring maintenance control systems', *International Journal of Operations & Production Management*, Vol. 14, No. 7, pp. 5-17.

Glazier, JD, & Powell, RR 1992, *Qualitative research in information management*, Libraries Unlimited Inc., Englewood.

Goetsch, DL 1990, *Advanced Manufacturing Technology*, Delmar Publisher Inc., NY.

Goetsch, DL 1996, *Occupational Safety and Health*, Prentice-Hall, Englewood Cliffs, NJ.

Golden, W. & Powell, P 2000, 'Towards a definition of flexibility: in search of the holy grail?', *OMEGA: The International Journal of Management Science*, Vol. 28, No. 4, pp. 373-384.

Gomes, CF, Yasin, MM, & Lisboa, JV 2004, 'A literature review of manufacturing performance measures and measurement in an organizational context: a framework and direction for future research', *Journal of Manufacturing Technology Management*, Vol. 15, No. 6, pp. 511-530.

Gomes, CF, Yasin, MM, & Lisboa, JV 2006, 'Performance measurement practices in manufacturing firms: an empirical investigation', *Journal of Manufacturing Technology Management*, Vol. 17 No. 2, pp. 144-167.

Gomolski, G, Grigg, J, & Potter, K 2001, 'IT Spending and Staffing Survey Results', Gartner Group Strategic Analysis Report, *Gartner Group.*

Gonzalez, R, Gasco, J, & Llopis, J 2005, 'Information systems outsourcing success factors: a review and some results', *Information Management & Computer Security*, Vol. 13, No. 5, pp. 399-418.

Goodhue, DL 1995, 'Understanding user evaluations of information systems', *Management Science*, Vol. 41, No. 12, pp. 1827-1844.

Goodhue, DL, & Thompson, RL 1995, 'Task-technology fit and individual performance', *MIS Quarterly*, Vol. 19, No. 2, pp. 213-236.

Goodhue, DL, Wybo, MD, & Kirsch, LJ 1992, 'The impact of data integration on the cost and benefits', *MIS Quarterly*, Vol. 16, No.3, pp.293-311.

Gordon, SR, & Gordon, JR 2002, 'Organizational options for resolving the tension between IT departments and business units in the delivery of IT services', *Information Technology and People*, Vol. 15, No. 4, pp.286-305.

Gosain, S, Lee, Z, Im, I 1997, 'Topics of Interest in IS: Comparing Academic Journals with the Practitioner Press', Proceedings of the eighteenth International Conference on Information Systems, Atlanta, Georgia.

Gosselin, M 2005, 'An empirical study of performance measurement in manufacturing firms', *International Journal of Productivity and Performance Management*, Vol. 54 No. 5/6, pp. 419-437.

Gottschalk, P 2006, 'Information systems in value configurations, *Industrial Management and Data Systems*, Vol. 106, No. 7, pp. 1060-1070.

Gregoire, YM, Wade, JH, & Antia, K 2001, 'Resource redeployment in an ecommerce environment: a resource-based view', *American Marketing Association Conference*, Long Beach, CA.

Grembergen WV, & De Haes, S 2005, 'Measuring and Improving IT Governance Through the Balanced Scorecard', *Information Systems Control Journal, Vol. 2.* accessed online on 10th July 2006, available online at http://www.isaca.org/Template.cfm?Section=Home&CONTENTID=24171&TEMPLATE=/ContentManagement/ContentDisplay.cfm.

Grembergen, WV, & Bruggen, RV 2003, 'Measuring and improving corporate information technology through the balanced scorecard', *Electronic Journal of Information Systems Evaluation*, EJISC, vol 1, no. 1.

Griffith, TL, Sawyer, JE, & Neale, MA 2003, 'Virtualness and knowledge in teams: Managing the love triangle of organizations, individuals, and information technology', *MIS Quarterly*, Vol. 27, No. 2, 265-287.

Grint, K, & Woolgar, S 1997, *The Machine at Work: Technology, Work and Organization*, Polity Press, Cambridge.

Grover, V, & Segars, AH 2005, 'An empirical evaluation of stages of strategic information systems planning: patterns of process design and effectiveness', *Information and Management*, Vol. 42 No.5, pp.761-779.

Gruber, T 1991, 'The role of common ontology in achieving sharable, reusable knowledge bases', in *Principles of Knowledge Representation and Reasoning* , eds. JA Allen, R Fikes, E Sandewall, Morgan Kaufmann, San Mateo, CA.

Guba, EG, & Lincoln, YS 1989, *Fourth generation evaluation*, Sage, Newbury Park, CA.

Guilfoyle, AM 2000, 'The challenge and the promise: a critical analysis of prejudice in intergroup attribution research', PhD dissertation, Murdoch University, Perth.

Gummesson, E 2000, *Qualitative Methods in Management Research*, Sage, London

Gupta, A, & Whitehouse, FR 2001, 'Firms using advanced manufacturing technology management: an empirical analysis based on size, *Integrated Manufacturing Systems*, Vol. 12, No. 5, pp. 346-350.

Gwillim, D, Dovey, K, & Wieder, B 2005, 'The politics of post-implementation reviews', *Information Systems Journal*, Vol. 15, pp.307–319.

Haider, A, & Koronios, A 2003, 'Managing Engineering Assets: A Knowledge Based Approach through Information Quality', in Proceedings of 2003 International Business Information Management Conference, Cairo, Egypt, pp.443-452.

Haider, A, & Koronios, A 2004a, 'Converging monitoring information for integrated predictive maintenance', in Proceedings of 3rd International Conference on Vibration Engineering & Technology of Machinery (VETOMAC-3) & 4th Asia -Pacific Conference on System Integrity and Maintenance, New Delhi, India.

Haider, A, & Koronios, A 2004b, 'RFID for Integrated Condition Monitoring', in Proceedings of 5th International We-B Conference 2004, Perth, WA.

Haider, A, & Koronios, A 2005, 'ICT Based Asset Management Framework', in Proceedings of 8th International Conference on Enterprise Information Systems, *ICEIS*, Paphos, Cyprus, vol. 3, pp. 312-322.

Haider, A, Koronios, A, & Quirchmayr, G 2006, 'You Cannot Manage What You Cannot Measure: An Information Systems Based Asset Management Perspective', in Proceedings of Proceedings of Inaugural World Congress on Engineering Asset Management, eds J. Mathew, L Ma, A Tan & D Anderson, 11-14 July 2006, Gold Coast, Australia

Hajo AR 2006, 'Implementing BPM systems: The role of process orientation', *Business Process Management Journal*, Vol. 12, No. 4, pp. 389-409.

Hamel, G, & Prahalad, CK 1989, 'Strategic Intent', *Harvard Business Review*, Vol. 67, May-June, pp. 63-76.

Hanna, MD, & Newman, WR 1995, 'Operations and environment: an expanded focus for TQM', *International Journal of Quality and Reliability Management*, Vol. 12, No. 5, pp. 38-53.

Hansen, T, Jensen, JM, & Solgaard, HS 2004, 'Predicting online grocery buying intention: A comparison of the theory of reasoned action and the theory of planned behavior', *International Journal of Information Management*, 24(6), pp. 539-550.

Hasan, B, and Ali, JMH 2004, 'An Empirical Examination of a Model of Computer Learning Performance', *The Journal of Computer Information Systems*, Vol. 44, No. 4, pp. 27-34.

Hastings, NAJ 2000, 'Asset management and maintenance', Queensland University of Technology, Brisbane, Queensland.

Hawgood, J & Land, F 1988, 'A Multivalent Approach to Information Systems Assessment', in *Information Systems Assessment: Issues and Challenges*, eds N Bjorn-Andersen & GB Davis, North Holland, Amsterdam, pp.103-124.

Hayes, N, & Walsham, G 2000, 'Competing interpretations of computer-supported cooperative work in organizational contexts', *Organization*, vol. 7, no. 1, pp. 49-67.

Heemstra, FJ, & Kusters, RJ 2004, 'Defining ICT proposals', *The Journal of Enterprise Information Management*, Vol. 17, No. 4, pp. 258-268.

Heilbroner, R 1994, 'Do machines make history?', in *Does Technology Drives History?- The Dilemma of Technological Determinism*, ed. L Marx, MIT Press, Cambridge, MA, pp. 53-65.

Helfert, M 2001, 'Managing and Measuring Data Quality in Data Warehousing', in Proceedings of the World Multiconference on Systemics, Cybernetics and Informatics, Florida, Orlando, pp. 55-65.

Hemsworth, D, Sanchez-Rodriguez, C, & Bidgood, B 2005, Determining the impact of quality management practices and purchasing-related information systems on purchasing performance - A structural model, *The Journal of Enterprise Information Management*, Vol. 18, No. 2, pp. 169-194.

Henderson, JC, & Venkatraman, N 1992, 'Strategic Alignment: A Model for Organizational Transformation Through Information Technology', in *Transforming Organizations*, eds. TA Kochan & M Useem, Oxford University Press, Oxford.

Henderson, JC, & Venkatraman, N 1993, 'Strategic Alignment: Leveraging Information Technology for Transforming Organizations', *IBM Systems Journal*, 32(1), pp. 4-16.

Heng, MSH & de Moor, A 2003, 'From Habermas's communicative theory to practice on the internet', *Information Systems Journal*, Vol. 13, No. 4, pp. 331-352.

Henwood, F, & Hart, A 2003, 'Articulating gender in the context of ICTs in health care: The case of electronic patient records in the maternity services', *Critical Social Policy*, Vol. 23, No. 2, pp. 249-267.

Hepworth, P 1998, 'Weighing it up: a literature review for the balanced scorecard', *Journal of Management Development*, Vol. 17, No. 8, p. 559-563.

Hidding, G 2001, 'Sustaining strategic IT advantage in the information age: how strategy paradigms differ by speed', *Strategic Information Systems*, 10(3), pp. 201-222.

Hinds, PJ, & Bailey, DE 2003, 'Out of sight, out of sync: Understanding conflict in distributed teams', *Organization Science*, Vol. 14, No. 6, pp.615-632.

Hinks, J 1998, 'A conceptual model for the interrelationship between information technology and facilities management process capability', *Facilities*, Vol. 16, No. 9/10, pp. 233-245.

Hipkin, I 2001, 'Knowledge and IS implementation: case studies in physical asset management', *International Journal of Operations & Production Management*, Vol. 21, No. 10, pp.1358-1380.

Hirschheim, R, & Klein, HK 1989, 'Four Paradigms of Information Systems Development', *Communications of the ACM*, Vol. 32, No. 10, pp. 1199-1216.

Hirschheim, R, & Smithson, S 1999, 'Evaluation of Information Systems: a Critical Assessment', In *Beyond the IT Productivity Paradox*, eds. LP Willcocks & L Stephanie, John Wiley and Sons, London.

Hirschheim, R, Klein, H, & Lyytinen, K 1995, *Information Systems Development and Data Modelling: conceptual and philosophical foundations,* Cambridge University Press, Cambridge.

Hitt, LM, and Brynjolfsson, E 1996, 'Productivity, Business Profitability, and Consumer Surplus: Three Different Measures of Information Technology Value', *MIS Quarterly,* Vol. 20, No. 2, pp. 121-142.

Honkanen, T 2004, 'Modelling Industrial Maintenance Systems and the Effects of Automatic Condition Monitoring', Thesis, *Helsinki University of Technology,* Espoo, Finland.

Hoque, Z, & James, W 2000, 'Linking balanced scorecard measures to size and market factors: impact on organisational performance', *Journal of Management Accounting Research,* Vol. 12, No.1, pp.1-15.

Hoxmeier, JA, Nie, W, & Purvis, GT 2000, 'The impact of gender and experience on user confidence in electronic mail', *Journal of End User Computing,* 12(4), pp. 11-20.

Huang, C, Fisher, N, Spreadborough, A, & Suchocki, M 2003, 'Identify in the critical factors of IT innovation adoption and implementation within the construction industry', in Proceedings of Second International Conference on Construction in the 21st Century (CITC-II), Sustainablity and Innovation in Management and Technology, 10-12 December, Hong Kong.

Huber, GP 1991, 'Organizational Learning: The Contributing Processes and the Literatures', *Organization Science,* Vol. 2, No. 1, pp. 88-115.

Hudson, R, Ralph, H, & Waheed, U 1997, *Infrastructure Management: Integrating Design, Construction, Maintenance, Rehabilitation, and Renovation,* McGraw-Hill, NY.

Humphreys, PK, Lai, MK, & Sculli, D 2001, 'An inter-organizational information system for supply chain management', *International Journal of Production Economics,* Vol. 70, No. 3, pp. 245-255.

Hurwicz, M 1997, 'Take Your Data to the Cleaners', *BYTE,* January.

Husband, TM 1976, *Maintenance Management and Terotechnology,* Saxson House, Hampshire, England.

Huseyin, T 2005, 'Information Technology Relatedness, Knowledge Management Capability, and Performance of Multibusiness Firms', *MIS Quarterly,* Vol. 29, No. 2, pp. 311-335.

Hussey, J & Hussey, R 1997, *Business Research,* Macmillan Press Ltd, Basingstoke.

Hutton, RW 1996, 'Condition monitoring the way forward', in *Handbook of Condition Monitoring,* ed. BKN Rao , Elsevier Advanced Technology, Oxford, UK.

Ibrahim, D 2006, Microcontroller based applied digital control, *John Wiley & Sons,* Chichester, England.

Ihde, D 2002, 'How could we ever believe science is not political?', *Technology in Society,* Vol 24, No. 1-2, pp. 179-189.

IIMM 2006, 'International Infrastructure Management Manual', Association of Local Government Engineering NZ Inc, National Asset Management Steering Group, New Zealand, Thames, ISBN 0-473-10685-X.

Incite 2006, 'Optus Incite Version 3 Quick Reference Guide', Optusincite, Optus Australia.

Ingemansson, A, & Bolmsjo, GS 2004, 'Improved efficiency with production disturbance reduction in manufacturing systems based on discrete-event simulation', *Journal of Manufacturing Technology Management*, Vol. 15 , No. 3, pp. 267-279.

Inman, RA 2002, 'Implications of environmental management for operations management', *Production Planning and Control*, Vol. 13, No.1, pp.47-55.

Intentia, 2004, 'Enterprise Asset Management Benchmark Survey: How do you measure your maintenance performance?', *The Maintenance Journal*. February, pp.52 -70.

Irani, Z, & Love, PED 2001, 'Editorial - Information Systems Evaluation: Past, Present and Future', *European Journal of Information Systems*, vol. 10, pp.183-188.

Iskandar, BY, Kurokawa, S, & LeBlanc, LJ 2001, 'Adoption of electronic data interchange: The role of buyer-supplier relationships', *IEEE Transactions on Engineering Management*, Vol. 48, No. 4, pp.505-517.

Ittner, D, Larcker, DF, & Randall, T 2003, 'Performance implications of strategic performance measurement in financial services firms', *Accounting Organisation and Society*, Vol. 28, No.7/8, pp.715-741.

Ivari, J 1991, 'A paradigmatic analysis of contemporary schools of IS development', *European Journal of Information Systems*, Vol. 1, No. 4, pp. 249-272.

JSF 2007, F-35 Lightening II program, accessed online 4/01/2007 at http://www.jsf.mil

Jae-Nam, L, & Young-Gul, K 2005, 'Understanding outsourcing partnership: A comparison of three theoretical perspectives', *IEEE Transactions on Engineering Management*, Vol. 52, No. 1, pp.43-58.

Jagodzinski, P, Reid, FJM, Culverhouse, P, Parsons, R, & Phillips, I 2000, 'A Study of Electronics Engineering Design Teams', *Design Studies*, 21(4), pp. 375-402.

Janson, M, & Cecez-Kecmanovic, D 2005, 'Making sense of e-commerce as social action', *Information Technology and People*, Vol. 14, No. 4, pp. 311-343.

Jardine, A 1999, The evolution of reliability: How RCM developed as a viable maintenance approach, *The Reliability Handbook: From Downtime to Uptime – in no time*, ed. JD Campbell, Plant Engineering & Maintenance, Clifford/Elliot Publication, Ontario.

Jarke, M, Lenzerini, M, Vassiliou, Y, & Vassiliadis, P 2003, *Fundamentals of DataWarehouses*, 2nd edition, Springer-Verlag, Berlin.

Jarvenpaa, SL 1988, 'The Importance of Laboratory Experimentation in Information Systems Research', *Communication so ACM*, Vol. 31, No. 12, pp. 1502-1504.

Jarvenpaa, SL, Shaw, TR, & Staples, DS 2004, 'Toward Contextualized Theories of Trust: The Role of Trust in Global Virtual Teams', *Information Systems Research*, Vol. 15, No. 3, pp. 250-267.

Jarvinen, J, Vannas, V, Mattila, M, & Larwowski, W 1996, 'Causes and Safety Effects of Production Disturbances in FMS Installations: A Comparison of Field Survey in the USA and Finland', *International Journal of Human Factors in Manufacturing*, Vol. 6, No. 1, pp.57-73.

Jaska, PV & Hogan, PT 2006, 'Effective management of the information technology function', *Management Research News*, Vol. 29, No. 8, pp. 464-470.

Jasperson, J, Carter, PE & Zmud, RW 2005, 'A Comprehensive Conceptualization of Post-Adoptive Behaviors Associated with Information Technology Enabled Work Systems', *MIS Quarterly*, Vol. 29, No. 3, pp. 525-557.

Jones, M, & Karsten, H 2003, 'Review: Structuration theory and Information Systems Research'. WP 11/03. Judge Institute Working Papers, University of Cambridge, 87 pages, online accessed on December 3, 2005 at http://www.jbs.cam.ac.uk/research/working_papers/2003/wp0311.pdf

Jonsson, P 1999, 'The Impact of Maintenance on the Production Process - Achieving High Performance', PhD Thesis, Division of Production Management, Lund University, Lund, Sweden.

Kallinikos, J 2006, 'Information out of information: on the self-referential dynamics of information growth', *Information Technology & People*, Vol. 19, No. 1, pp. 98-115.

Kamal, MM 2006, IT innovation adoption in the government sector: identifying the critical success factors', *Journal of Enterprise Information Management*, Vol. 19, No. 2, pp. 192-222.

Kaplan, B & Maxwell, JA 1994, 'Qualitative Research Methods for Evaluating Computer Information Systems', in *Evaluating Health Care Information Systems: Methods and Applications,* eds JG Anderson, CE Aydin & SJ Jay, Sage, Thousand Oaks, CA.

Kaplan, B, & Duchon, D 1988, 'Combining Qualitative and Quantitative Methods in Information Systems Research: A Case Study', *MIS Quarterly*, Vol. 12, No. 4, pp. 571-587.

Kaplan, RS, & Norton, DP 1992, 'The balanced scorecard – measures that drive performance', *Harvard Business Review*, Vol. 70 No. 1, pp. 71-79.

Kaplan, RS, & Norton, DP 1996, *The Balanced Scorecard: Translating Strategy into Action*, Harvard Business School Press, Cambridge, MA.

Kaplan, RS, & Norton, DP 2001, 'Transforming the Balanced Scorecard From Performance Measurement to Strategic Management: Part II', *Accounting Horizons*, Vol. 15, No. 2, pp. 147-160.

Kaplan, RS, & Norton, DP, 2004, 'The strategy map: guide to aligning intangible assets', *Strategy & Leadership*, Vol. 32, No. 5, pp. 10-17.

Kappelman, LA, & Mclean, ER 1994, 'User Engagement in Information Systems Development', in *Diffusion, Transfer and Implementation of Information Technology,* ed. L. Levine, Elsevier Science, Amsterdam.

Karababas, S, & Cather, H 1994, 'Developing Strategic Information Systems', *Integrated Manufacturing Systems,* Vol. 5, No. 2, pp. 4-11.

Karlsson, T, & Gennas, JB 2005, 'Content Management Systems – Business effects of an implementation', Master Thesis in Informatics, Department of Informatics, Goteborg University and Chalmers University of Technology, Goteborg, Sweden.

Kathuria, R, & Porth, SJ 2003, 'Strategy-managerial characteristics alignment and performance: A manufacturing perspective', *International Journal of Operations & Production Management*, Vol. 23, No. 3, pp. 255-276.

Kauffman, RJ, & Mohtadi, H 2004, 'Proprietary and open systems adoption in E-procurement: A risk-augmented transaction cost perspective, *Journal of Management Information Systems*, Vol. 21, No. 1, pp. 137-166.

Keil, M, Smith, HJ, Pawlowski, S, & Jin, L 2004, 'Why didn't somebody tell me?: Climate, information asymmetry, and bad news about troubled projects', *Database for Advances in Information Systems*, Vol 35, No. 2, pp. 65-84.

Kelly, A 1989, *Maintenance and its Management*, Ashford Press Ltd, Southampton, Hampshire.

Kennerley, M, & Neely, A 2002, 'A framework of the factors affecting the evolution of performance measurement systems', *International Journal of Operations & Production Management*, Vol. 22, No.11, pp.1222-1245.

Kern, T, Kreijger, J, & Willcocks, L 2002, 'Exploring ASP as sourcing strategy: Theoretical perspectives, propositions for practice', *Journal of Strategic Information Systems*, Vol. 11, No. 2, pp. 153-177.

Kerns, F 1999, 'Strategic facility planning (SFP)', *Work Study*, 48(5), pp. 176-181.

Kerr, RM, & Greenhalgh, GR 1991, 'Aspects of manufacturing strategy', *Production Planning and Control*, Vol. 2, No.3, pp.194-206.

Khalifa, G, Irani, Z, Baldwin, LP, & Jones, S 2001, 'Evaluating Information Technology with You in Mind', *Electronic Journal of Information Systems Evaluation*, EJISE, vol. 4, issue 1.

Khazanchi, D 2005, 'Information technology (IT) appropriateness: The contingency theory of fit and IT implementation in small and medium enterprises', *Journal of Computer Information Systems*, Vol. 45, No. 3, pp.88-95.

Kim, KK & Michelman, JE 1990, 'An examination of factors for the strategic use of information systems in the health care industry', *MIS Quarterly*, Vol. 14, No. 2, pp. 201-215.

King, AA, & Lenox, MJ 2001, 'Lean and green? An empirical examination of the relationship between lean production and environmental performance', *Production and Operations Management*, Vol. 10 No.3, pp.244-56.

King, M, & McAulay, L 1997, 'Information technology investment evaluation: evidence and interpretations, *Journal of Information Technology*, Vol. 12, No. 2, pp. 131-143.

Kirkpatrick, I, & Lucio, MM 1995, *The Politics of Quality in the Public Sector*, Routledge, London.

Klecun, E, & Cornford, T 2005, 'A critical approach to evaluation', *European journal of information systems*, 14 (3). pp. 229-243.

Klein, HK & Myers, MD 1999, 'A set of principles for conducting and evaluating interpretive field studies in information systems', *MIS Quarterly*, 23(1), pp. 67-93.

Kling, R, McKim , G, & King, A 2003, 'A bit more to it: Scholarly communication forums as socio- technical interaction networks', *Journal of the American Society for Information Science and Technology*, Vol. 54, No. 1, pp.47-67.

Knezevic, J 1992, *Reliability, Maintainability, Supportability, A Probabilistic Approach*, McGraw-Hill, London.

Knezevic, J 1995, 'Planning maintenance resources for non-supported missions', *Journal of Quality in Maintenance Engineering*, Vol. 1, No. 1, pp. 60-68.

Ko, D, Kirsch, LJ, & King, WR 2005, 'Antecedents of knowledge transfer from consultants to clients in enterprise system implementations', *MIS Quarterly*, Vol. 29, No. 1, pp. 59-85.

Kochhar, AK, Howard, AP, & Kilworth, J 1998, 'Manufacturing planning and control systems in virtual enterprises – a rule-based system for convergence assessment', *Proceedings of XI International Production Technology Conference*, pp.297-303.

Kohli, R, & Kettinger, WJ 2004, 'Informating the clan: Controlling physicians' costs and Outcomes', *MIS Quarterly*, Vol. 28, No. 3, pp.363 -394.

Konradt, U, Zimolong, B, & Majonica, B 1998, 'User-Centred Software Development: Methodology and Usability Issues', in *The Occupational Ergonomics Handbook*, eds W Karwowski, & WS Marras, USA, CRC Press.

Konsynski, BR 1993, 'Strategic control in the extended enterprise', *IBM Systems Journal*, Vol. 32 No.1, pp.111-42.

Kotha, S, & Swamidass, PM 2000, 'Strategy, advanced manufacturing technology and performance: empirical evidence from US manufacturing firms', *Journal of Operations Management*, Vol. 18, pp. 257-277.

Kraft, P, & Truex, D 1994, 'Postmodern Management and Information Technology in the Modern Industrial Corporation', in Proceedings of the IFIP WG8.2 Working Conference on Information Technology and New Emergent Forms of Organization, Ann Arbor, Michigan, USA, 11-13 August, 1994, eds. R Baskerville S Smithson O Ngwenyama & J DeGross, North-Holland, New York.

Kuivanen, R 1996, Disturbance control in flexible manufacturing, *The International Journal of Human Factors in Manufacturing*, Vol. 6, No. 1, pp.41-56.

Kulak, O, Kahraman, C, Oztaysi,B, & Tanyaş, M 2005, 'Multi-attribute information technology project selection using fuzzy axiomatic design', *The Journal of Enterprise Information Management*, Vol. 18, No. 3, pp. 275-288.

Kumar, K 1990, 'Post Implementation Evaluation of Computer-Based Information Systems: Current Practices', *Communications of the ACM*, 33(2), pp. 203-212.

Kumar, S, & Harms, R 2004,'Improving business processes for increased operational efficiency: a case study', *Journal of Manufacturing Technology Management*, Vol. 15, No. 7, pp. 662–674.

Kumar, UD, & Knezevic, J 1998, 'Supportability - critical factor on systems' operational availability', *International Journal of Quality & Reliability Management*, Vol. 15, No. 4, pp. 366-376.

Kunnathur, AS, & Shi, Z 2001, 'An investigation of the strategic information systems planning success in Chinese publicly traded firms', *International Journal of Information Management*, Vol. 21, No.6, pp.423-439.

Kuo, FY, Chu, TH, Hsu, MH & Hsieh, HS 2004, 'An investigation of effort-accuracy trade-off and the impact of self-efficacy on Web searching behaviors', *Decision Support Systems*, Vol. 37, No. 3, pp. 331-342.

Kurnia, S, Johnston, RB 2000, 'The need for a processual view of inter-organizational systems adoption', *Journal of Strategic Information Systems*, Vol. 9 pp.295-319.

Kutucuoglu, KY, Hamali, J, Irani, Z, & Sharp, JM 2001, 'A framework for managing maintenance using performance measurement systems, *International Journal of Operations & Production Management*, Vol. 21, No. 1/2, pp. 173-194.

Kuwaiti, ME 2004, 'Performance measurement process: definition and ownership', *International Journal of Operations & Production Management*, 24(1), pp. 55-78.

Kwon, TH, & Zmud, RW 1987, 'Unifying the Fragmented Models of Information Systems Implementation', in *Critical Issues in Information Systems Research*, eds. RJ Boland Jr. & RA Hirshheim, John Wiley & Sons, NY.

Lai, F, Zhao, X, & Wang, Q 2006, 'The impact of information technology on the competitive advantage of logistics firms in China', *Industrial Management & Data Systems*, Vol. 106, No. 9, pp. 1249-1271.

Lamb, R, & Kling, R 2003, 'Reconceptualizing users as social actors in Information Systems Research', *MIS Quarterly*, Vol.27, No. 2, pp. 197-235.

Land, F 1992, 'The information systems domain', in *Information Systems Research: issues, methodologies and practical guidelines*, ed. RD Galliers, Blackwell Scientific, Oxford, pp 6-13.

Langfiled-Smith, K, Thorne, H, & Hilton, RW 2006, *Management Accounting: Information for managing and creating value*, 4th edition, McGraw Hill, Sydney, NSW.

Lankhorst, M (2005), Enterprise Architecture at Work: Modeling, Communication and Analysis, Springer-Verlag Berlin Heidelberg, Germany

Lapiedra, R, Alegre, J & Chiva, R 2006, 'User participation on the development of information systems', in *European and Mediterranean Conference on Information Systems (EMCIS)*, Costa Blanca, Alicante, Spain, pp. 1-10

Larsen, T, Levine, L, & DeGross, JI (eds) 1999, 'Information systems: current issues and future changes', Laxenburg: IFIP.

Lau, HCW, Lau, PKH, Chan, FTS, & Ip, RWL, 2001, 'A real time performance measurement technique for dispersed manufacturing', *Measuring Business Excellence*, Vol. 5. No. 1, pp. 9-15.

Laudon, K, & Laudon, J 1998, *Management Information Systems New Approaches to Organization & Technology*, 7th edition, Prentice Hall International In., Upper Saddle, New Jersey.

Laurindo, FJB, & de Carvalho, MM 2005, 'Changing product development process through information technology: a Brazilian case', *Journal of Manufacturing Technology Management*, Vol. 16 No. 3, pp. 312-327.

Lebas, MJ 1995, 'Performance Measurement and Performance Management', *International Journal of Production Economics*, Vol. 41, No. 1-3, pp. 23-35.

Lederer, AL, & Sethi V, 1996, 'Key Prescriptions for Strategic Information Systems Planning', *Journal of Management Information Systems*, Vol. 13, No. 1, pp. 35-62.

Ledington, PWJ & Ledington, J 1999, 'The problem of comparison in soft systems methodology', *Systems Research and Behavioral* Science, 16(4), pp. 329-339.

Lee, AS 1994, 'Integrating Positivist and Interpretive Approaches to Organizational Research', *Organization Science*, Vol. 2, No. 4, pp. 342-365.

Lee, AS, & Liebenau, J 1997, 'Information systems and qualitative research', in *Information systems and qualitative research*, eds. AS Lee J Liebenau & JI DeGross, Chapman and Hall, London, pp. 1-8.

Lee, J, Ni, J, & Koc, M 2001, 'NSF Workshop on Tether-free Technologies for e-Manufacturing, e-Maintenance & e-Service', IMS: Centre for Intelligent Maintenance Systems, Milwaukee, accessed online at http://www.www.uwm.edu/CEAS/ims/PDF2/WorkshopReport.pdf,on 25/09/2005.

Lee, GG, & Pai, RJ 2003, 'Effects of organizational context and inter-group behaviour on the success of strategic information systems planning: an empirical study', *Behaviour and Information Technology*, Vol. 22 No.4, pp.263-280.

Lee, I 2004, 'Evaluating business process-integrated information technology investment', *Business Process Management Journal*, Vol. 10, No. 2, pp. 214-233.

Lee, WB, Cheung, HCW, Choy, L, & Choy, KL 2003, 'Development of a web-based enterprise collaborative platform for networked enterprises', *Business Process Management Journal*, Vol. 9 No.1, pp.46-58.

Lei, D, Hitt, M, & Goldhar, J 1996, 'Advanced manufacturing technology: organizational design and strategic flexibility', *Organization Studies*, Vol. 17 No.3, pp.501-523.

Leibs, S 2002, 'A step ahead: Economist Erik Brynjolfsson leads the charge toward a greater appreciation of IT', *CFO Magazine*, NY, pp.38-41.

Leonard, LNK, Cronan, TP, & Kreie, J 2004, 'What influences IT ethical behavior intentions-planned behavior, reasoned action, perceived importance, or individual characteristics?', *Information and Management*, Vol. 42, No. 1, pp. 143-158.

Lesjak, D, & Vehovar, V 2005, 'Factors affecting evaluation of e-business projects', *Industrial Management & Data Systems*, Vol. 105, No. 4 , pp. 409-428.

Letza, S 1996, 'The Design and Implementation of the Balanced Business Scorecard – an Analysis of Three Companies in Practice', *Business Process Management Journal*, Vol. 2, No. 3, pp. 54-76.

Leveson, N 2005, 'Safety in Integrated Systems Health Engineering and Management, In Proceedings of NASA Ames Integrated System Health Engineering and Management Forum (ISHEM), Napa.

Levitan, AV, & Redman, TC 1998, 'Data as a resource: properties, implications and prescriptions', *Sloan Management Review*, Vol. 40, No.1, pp.89-101.

Lewin, D 1995, 'Predictive maintenance using PCA', Control Engineering Practice, Vol. 3, No. 3, pp. 415-421.

Liang, P, Song, F 1994, 'Computer-Aided Risk Evaluation System for Capital Investment', *International Journal of Management Science*, 22(4), pp.391-400.

Liaw, SS, Chang, WC, Hung, WH, & Huang, HM 2006, 'Attitudes toward search engines as a learning assisted tool: Approach of Liaw and Huang's research model', *Computers in Human Behavior*, Vol. 22, No. 2, pp. 177-190.

Lim, K, Benbasat, I 2000, 'The Effect of Multimedia on Perceived Equivocality and Perceived Usefulness of Information Systems', *MIS Quarterly*, 24(3), pp. 449-471.

Lin, C, & Pervan, G 2001, 'A review of IS/IT investment evaluation and benefits management issues, problems, and processes', In *Information Technology Evaluation Methods and Management,* ed. WV Grembergen, John Wiley & Sons, New York, NY.

Lin, HF, & Lee, GG 2004, 'Perceptions of senior managers toward knowledge-sharing behaviour', *Management Decision*, Vol. 42 No.1, pp.108-25.

Lincoln, YS, & Guba EG 1985, *Naturalistic Inquiry*, Sage Publications, London, U.K.

Lindberg, P 1990, 'Strategic Manufacturing Management: A Proactive Approach', *International Journal of Operations and Production Management*, 10(2), pp. 94-106.

Liyanage, JP, & Kumar, U 2000, 'Utility of maintenance performance indicators in consolidating technical and operational health beyond the regulatory compliance', in Proceedings of Safety Engineering and Risk Analysis: The International Mechanical Engineering Congress and Exposition-2000, pp.153-160.

Liyanage, JP, & Kumar, U 2003, 'Towards a value-based view on operations and maintenance performance management', *Journal of Quality in Maintenance Engineering*, vol. 9, no. 4, pp. 333-350

Llorens-Montes, F, Garcia-Morales, V, & Verdu-Jover, A 2004, 'Flexibility and quality management in manufacturing: an alternative approach', *Production Planning and Control*, Vol. 15, No.5, pp.525-33.

Loch, CH & Huberman, BA 1999, 'A punctuated equilibrium model of technology diffusion', *Management Science*, Vol. 45, No. 2, pp. 160-177.

Lockamy, A 1998, 'Quality-focused performance measurement systems: a normative model', *International Journal of Operations and Production Management*, Vol. 18, No. 8, pp. 740-766.

Love, PED, Irani, Z, Li, H, Cheng, EWL, & Tse, RYC 2001, 'An empirical analysis of the barriers to implementing e-commerce in small-medium sized construction contractors in the state of Victoria, Australia', *Construction Innovation*, Vol. 1, No. 1, pp. 31-41.

Lubbe, S, & Remenyi, D 1999, 'Management of information technology evaluation - the development of a managerial thesis', *Logistics Information Management*, Vol. 12, No. 1/2, pp. 145-156.

Lucas, H 1989, *Managing Information Services*, Macmillan Publishing Company, London.

Luna-Reyes, LF, Zhang, J, Gil-Garcia, JR & Cresswell, AM 2005, 'Information systems development as emergent socio-technical change: a practice approach', *European Journal of Information Systems*, vol. 14, no. 1, pp. 93-105

Lycke, L 2000, 'Implementing Total Productive Maintenance: driving forces and obstacles', Licentiate thesis. Department of Business Administration and Social Science, Lulea University, Sweden.

Lyytinen, KJ, & Klein, HK 1985, 'The critical theory of Jurgen Habermas as a basis for a theory of information systems', in *Research methods in information systems*, eds. E Mumford R Hirsschein G Fitzgerald & AT Wood-Harper, Amsterdam, Elsevier Science Publishing Company Inc., North-Holland, pp. 219-236.

Lyytinen, K 1992, 'Information systems and critical theory', in *Critical Management Studies,* eds. M Alvesson & H Willmott, Sage Publications, London, pp. 159-180.

MacKenzie, DA, & Wajcman, J (eds.) 1999, *The Social Shaping of Technology*, 2nd edition, Open University Press, Buckingham.

Madey, G, Freeh, V, & Tynan, R 2002, 'The Open Source Software Development Phenomenon: An Analysis Based On Social Network Theory', in Proceedimgs of Americas Conference on Information Systems (AMCIS2002), Dallas, TX.

Mahaney, RC, & Lederer, AL 2003, 'Information systems project management: An agency theory interpretation. *The Journal of Systems and Software*, 68(1), pp. 1-9.

Mahmood, M 1990, 'Information systems implementation success: A causal analysis using the linear structural relations model', *Information Resources Management Journal*, Fall, pp. 2-14.

Mahmood, MA, & Mann, GJ 1993, 'Measuring the organizational impact of information technology investment: An exploratory study', *Journal of Management Information Systems*, Vol. 10, No. 1, pp. 97-122.

Mahoney, LS, Roush, PB, & Bandy, D 2003, 'An Investigation of the effects of decisional guidance and cognitive ability on decision-making involving uncertainty data', *Information and Organization*, Vol. 13, No. 2, pp. 85-110.

Majchrzak, A, Malhotra, A, & John, R 2005, 'Perceived individual collaboration know-how development through information technology-enabled contextualization: Evidence from distributed teams', *Information Systems Research*, 16(1), pp. 9-27.

Malhotra, A, Gosain, S, & El Sawy, OA 2005, 'Absorptive Capacity Configurations in Supply Chains: Gearing for Partner-Enabled Market Knowledge Creation', MIS *Quarterly*, Vol. 29, No. 1, pp. 145-187.

Mandeville, T 1998, 'An information economics perspective on innovation', *International Journal of Social Economics*, Vol. 25, No. 2/3/4, pp. 357-364.

Manion, M, & Evan WM 2002, 'Technological catastrophes: their causes and prevention'. *Technology in Society*, vol. 24, pp. 207-224.

Manoochehri, G 1999, 'The road to manufacturing excellence: using performance measures to become world-class', *Industrial Management*, March/April, pp. 7-13.

Mapes, J, Szwejczewski, M, & New, C 2000, 'Process variability and its effect on plant performance', *International Journal of Operations & Production Management*, Vol. 20 No. 7, pp. 792-808.

Markeset, T, & Kumar, U 2005, 'Product support strategy: conventional versus functional products', *Journal of Quality in Maintenance Engineering*, Vol. 11, No. 1, pp. 53-67.

Markus, ML, & Robey, D 1988, 'Information Technology and Organizational Change: Casual Structuring in theory and research', *Management Science*, 34(5), pp. 583-598.

Markus, M, Tanis, C, & Fenema, P 2000, 'Multi-site ERP implementation', *Communications of the ACM*, Vol. 43, No.4, pp.42-46.

Markus, ML, Majchrzak, A, & Gasser, L 2002, 'A design theory for systems that support emergent knowledge processes', *MIS Quarterly*, Vol. 26, No. 3, pp. 179-212.

Marosszeky, M, Sauer, C, Johnson, K, Karim, K, & Yetton, P 2000, 'Information Technology in the Building and Construction Industry: The Australian Experience', in Proceedings of the INCITE 2000 Conference, Implementing IT To Obtain a Competitive Advantage in the 21st Century, The Hong Kong Polytechnic University, Hong Kong, eds. H Li, Q Shen, D Scott and PED Love, Hong Kong Polytechnic University press: pp.78-92.

Marr, B, Gray, D, and Neely, A 2003, 'Why do firms measure their intellectual capital?', *Journal of Intellectual Capital*, Vol.. 4 No. 4, pp. 441 – 464.

Marsh, L, & Flanagan, R 2000, 'Measuring the costs and benefits of information technology in construction', *Engineering Construction & Architectural Management*, Vol. 7, No. 4, pp. 423-435.

Martins, RA (2000), 'Use of Performance Measurement Systems: Some Thoughts Towards a Comprehensive Approach', in Proceedings of Performance Measurement - Past, Present and Future, Cambridge, Centre for Business Performance, UK.

Martins, RA 2002, 'The Use of Performance Measurement Information as a Driver in Designing a Performance Measurement System', in *Performance Measurement and Management: Research and Action*, eds. A Neely, A Walters, & R Austin, Centre for Business Performance, Cranfield Business School, UK,

Marx, K 1847, *The Poverty of Philosophy*, accessed online on October 21, 2006, at http://www.marxists.org/archive/marx/works/1847/poverty-philosophy/ch02.htm.

Maskell, BH, *Performance Measurement for World Class Manufacturing*, Productivity Press, Cambridge, MA.

Massey, AP, & Montoya-Weiss, MM 2006, 'Unraveling the temporal fabric of knowledge conversion: A model of media selection and use', *MIS Quarterly*, 30(1), pp. 99-114.

Matson, JB, & McFarlane, DC 1999, 'Assessing the responsiveness of existing production operations', *International Journal of Operations & Production Management*, Vol. 19, No. 8, pp. 765-784.

McAdam, R 2000, 'Quality models in an SME context', *International Journal of Quality and Reliability Management*, Vol. 17, pp. 305–323.

McMaster, TE, Mumford, EB, Swanson, EB, Warboys, B, & Wastell, D (eds.) 1997, *Facilitating technology transfer through partnership: Learning from practice and research*, Chapman and Hall, London.

McNurlin, BC, & Sprague RH Jr. 2002, '*Information Systems Management in Practice*', Fifth Edition, Prentice Hall, Upper Saddle River, NJ.

Medori, D, & Steeple, D 2000, 'A framework for auditing and enhancing performance measurement systems', *International Journal of Operations & Production Management*, Vol. 20, No. 5, pp. 520-533.

Meekings, A 1995, 'Unlocking the potential of performance measurement: A practical implementation guide', *Public Money and Management*, Vol. 15, No. 4, pp. 5-12.

Melville, N, Kraemer, KL, & Gurbaxani, V 2004, 'Information technology and organizational performance: An integrative model of IT business value', *MIS Quarterly*, Vol. 28, No. 2, pp. 283-322.

Metheny, B 1994, 'Relying solely on productivity to measure the impact of information technology is measleading, says National Research Council study', *Journal of Systems Management*, Vol. 45, No. 3, pp. 24.

MGI (2001), 'U.S. Productivity Growth 1995-2000: Understanding the Contribution of Information Technology Relative to Other Factors' *McKinsey Global Institute*, High Tech Practice, Business technology Office, San Francisco, US.

MGI (2002) ' How IT enables productivity Growth: The US experience across three sectors in 1990s", *McKinsey Global Institute*, High Tech Practice, Business technology Office, San Francisco, US.

Miles, MB, & Huberman, AM 1994, *Qualitative data analysis: An expanded sourcebook*, 2nd edition, Sage publications, Thousand Oaks, CA.

Miles, BL & Swift, K 1998, 'Design for manufacture and assembly', Manufacturing Engineer, Vol. 77, No. 5, pp. 221 - 224.

Mintzberg, H 1990, 'The Design School: reconsidering the basic premises of strategic management', *Strategic Management Journal*, Vol. 11, No. 3.

Mirchandani, DA & Lederer, AL 2004, 'IS planning autonomy in US subsidiaries of multinational firms', *Information and Management*, Vol. 41, No. 8, pp. 1021-1036.

Mitchell, JS, & Carlson, J 2001, 'Equipment asset management – what are the real requirements?', *Reliability Magazine*, October, pp. 4-14.

Mora, M, Gelman, O, Cervantes, F, Mejia, M, & Weitzenfeld, A 2003, 'A systemic approach for the formalization of the information systems concept: why information systems are systems?', in *Critical Reflections on information Systems: A Systemic Approach*, ed. JJ Cano, Idea Group Publishing, Hershey, PA.

Moree (2001), 'Moree Rural Roads Funding Report', Moree Rural Roads Steering Group, accessed online on November 26, 2006 at http://www.algin.net/transinfo/Butcher%20report.pdf.

Morse, J, & Fields, PA 1995, *Qualitative Research Methods for Health Professionals*, Sage, Thousand Oaks, CA.

Moubray, J 1997, *Reliability-centered maintenance*, 2nd edition, Industrial Press Inc., NY.

Moubray, JM 2000, *Reliability-centred Maintenance*, Butterworth-Heinemann, Oxford.

Moubray, JM 2003, 21st Century Maintenance Organization Part I: The Asset Management Model, Maintenance Technology, accessed online at http://www.mt-online.com/articles/0203_asset_mgmtt.cfm, on July 12, 2006.

Mukherji, A 2002, 'The evolution of information systems: their impact on organizations and structures, *Management Decision*, Vol. 40, No. 5, pp. 497-507.

Mumford, E, & Weir, M 1979, *Computer Systems in Work Design: the ETHICS method*, Associated Business Press, London.

Mumford, E 1983, *Designing Human Systems*, Manchester Business School Publications, Manchester, UK.

Mumford, E 1995, 'Creative chaos or constructive change: business process re-engineering versus socio-technical design?', in *Examining Business Process Re-engineering: Current Perspectives and Research Directions*, eds. G Burke & J Peppard, Kogan Page, London, pp.192-216.

Mumford, E 1996, *Systems Design: ethical tools for ethical change*, Macmillan, London.

Mumford, E 2000, Socio-technical design: An Unfulfilled Promise or a future Opportunity, in *Organizational and Social Perspectives on Information Technology*, eds. R Baskerville, J Stage, & JI DeGross, Boston, Kluwer academic Publications.

Murthy, DNP, Atrens, A, & Eccleston, JA 2002, 'Strategic maintenance management', *Journal of Quality in Maintenance Engineering*, Vol. 8, No. 4, pp. 287-305.

Myers, BL, Kappelman, LA, & Prybutok, VR 1997, 'A comprehensive model for assessing the quality and productivity of the information systems function: toward a theory for information systems assessment', *Information Resources Management Journal*, Vol. 10, Issue 1, Winter, pp. 6-25.

Myers, MD 1997, 'Interpretive research in information systems, in *Information systems: An emerging discipline?*, eds. J Mingers & F Stowell, McGraw Hill, London.

Myers, MD 1999, 'Investigating Information Systems with Ethnographic Research', *Communication of the AIS*, Vol. 2, No. 23, pp. 1-20.

Nachiappan, RM, & Anantharaman, N 2006, 'Evaluation of overall line effectiveness (OLE) in a continuous product line manufacturing system, *Journal of Manufacturing Technology Management*, Vol. 17 No. 7, pp. 987-1008.

Nakajima, S 1988, *Introduction to TPM*. Productivity Press, Portland, OR.

Narain, R, Yadav, R, Sarkis, J, & Cordeiro, J 2000, 'The strategic implications of flexibility in manufacturing systems', *International Journal of Agile Management Systems*, Vol. 2, No.3, pp.202-13.

Naumann, F, & Rolker, C 2000, 'Assessment Methods for Information Quality Criteria", in Proceedings of the 2000 Conference on Information Quality, Cambridge, MA.

Neely, A, Mills, J, Gregory, M, & Platts, K 1995, 'Performance measurement system design – A literature review and research agenda', *International Journal of Operations & Production Management*, Vol. 15, No.4, pp.80-116.

Neely, AD, Gregory, MJ, & Platts, K 1995, 'Performance Measurement System Design: A Literature Review and Research Agenda', *International Journal of Operations & Production Management*, Vol. 15, No. 4, pp. 80-116.

Neely, A, Mills, J, Gregory, M, Richards, H, Platts, K, & Bourne, M 1996, *Getting the Measure of Your Business*. Findlay, London.

Neely, AD, Mills, JF, Gregory, MJ, Richards, AH, Platts, KW, & Bourn, MCS 1996, *Getting the Measure of your Business*, Findlay Publications, London.

Neely, A, Richards, H, Mills, J, Platts, K, & Bourne, M 1997, 'Designing Performance Measures: a Structured Approach', *International Journal of Operations and Production Management*, Vol. 17, No. 11, pp. 1131-1152.

Neely, A, Adams, C 2001, 'The performance prism perspective', Journal of Cost Management, Vol. 15, No. 1, pp.7-15.

Neely, A, Adams, C, & Crowe, P 2001, 'The performance prism in practice', *Measuring Business Excellence*, Vol. 5, No. 2, pp.6-12.

Neely, A, Adams, C, & Kennerley, M 2002, *The Performance Prism: The Scorecard for Measuring and Managing Business Success*, Financial Times-Prentice Hall: London.

Neely A, Adams C, & Kennerley M 2002, *The Performance Prism: The Scorecard for Measuring and Managing Business Success*, Financial Times Prentice Hall, NY.

Neely, A, Kennerley, M, & Martinez, V 2004, 'Does the balanced scorecard work: an empirical investigation', in Proceedings of Performance Measurement Association Conference, Edinburgh, July.

Neely, AD 1998, *Measuring Business Performance: Why, What and How*, Economist Books, London.

Nelson MR 2001, 'Alignment through cross-functional integration', in *Strategic information technology: opportunities for competitive advantage,* ed. R Papp, Idea Group Publishing, Hershey, PA, pp 40–55.

Nevis, EC, DiBeila, AJ & Gould, JM 1995, 'Understanding organizations as learning systems', *Sloan Management Review*, Winter, pp. 73-85.

Newkirk, HE, Lederer, AL, & Srinivasan, C 2003, 'Strategic information systems planning: too little or too much', *Journal of Strategic Information Systems*, Vol. 12, No.3, pp.201-228.

Newman, M & Robey, D 1992, 'A social process model of user-analyst relationships', *MIS Quarterly*, Vol. 16, No. 2, pp. 249-266.

Niebel, BW 1996, *Engineering Maintenance Management*, 2nd edn., Marcel Dekker, NY.

Nijland, MH 2004, 'Understanding the Use of IT Evaluation Methods in Organisations', PhD Thesis, Department of Information Systems, London School of Economics and Political Science, University of London, London.

Nitithamyong, P, & Skibniewski, MJ 2004, 'Web-based construction project management systems: how to make them successful?', *Automation in Construction*, Vol. 13, No. 4, pp. 491-506.

Nonaka, I 1991, 'Knowledge-creating company', *Harvard Business Review*, vol. 69, no. 6, pp.96-104.

Nonaka, I, & Takeuchi, H 1995, *The Knowledge-Creating Company*, Oxford University Press, New York, NY.

NSWG 2001, 'Asset disposal: strategic planning', New South Wales Government Asset Management Committee, DPWS Rpt No. 01048, ISBN 0734741405, Sydney, NSW.

Oakland, JS 1995, *Total Quality Management: Text with Cases*, Butterworth-Heinemann, New York, NY.

OALD 2005, 'The Oxford Advanced Learner's Dictionary', 7th revised edition, Oxford University Press, ISBN: 0194316491.

O'Brien, JA, & Morgan, JN 1991, 'A multidimensional model of information resource management', *Information Resources Management Journal,* Vol. 4, No. 2, pp. 2-11.

O'Brien, WJ 2000, 'Implementation issues in project web sites: A practitioner's viewpoint', *Journal of Management in Engineering,* Vol. 16, No. 3, pp. 34-39.

OECD 2006, 'OECD Information Technology Outlook 2006', *OECD,* ISBN Number: 92-64-02643-6, Paris.

Organ, G 1997, *Images of Organization,* Thousand Oaks, Sage Publications, CA.

Orlikowski, W, & Robey, D 1991, 'Information Technology and the Structuring of Organizations', *Information Systems Research,* Vol. 2, No. 2, pp. 143-169.

Orlikowski, WJ and Baroudi JJ 1991, 'Studying Information Technology in Organizations: Research Approaches and Assumptions', *Information Systems Research,* Vol. 2, No. 1, pp. 1-28.

Orlikowski, WJ 1992, 'The Duality of Technology: Rethinking the Concept of Technology in Organizations', *Organization Science.* Vol.3, No.3, pp. 398-427.

Orlikowski, W 1993, 'CASE tools as organisational change: Investing incremental and radical changes in systems development', *MIS Quarterly,* 17(3), pp. 309-340.

Orlikowski, WJ 1996, 'Improvising organizational transformation over time: a situated change perspective', *Information Systems Research,* vol. 7, no. 1, pp.63-92.

Orlikowski, WJ, Walsham, G, Jones, M, & DeGross, JI (eds.) 1996, *Information technology and changes in organizational work,* Chapman and Hall, London.

Orlikowski, WJ 2000, 'Using technology and constituting structures: A practice lens for studying technology in organizations', *Organization Science,* 11(4), pp. 404-428

Orlikowski, WJ, & Barley, SR 2001, 'Technology and institutions: what can research on information technology and research on organizations learn from each other?', *MIS Quarterly,* Vol. 25, No. 2, pp. 245-265.

Orr, K 1998, 'Data Quality and Systems Theory', *Communications of the ACM,* Vol. 41, No. 2, pp. 66-71.

Otley, DT 1999, 'Performance Management: A Framework for Management Control Systems Research', *Management Accounting Research,* Vol. 10, No. 4, pp. 363-382.

Owens, I, & Davies, PB 1999, 'The post-implementation evaluation of mission-critical information systems', 7th European Conference of Information Systems, June 23-25, Copenhagen, Denmark.

Paiva, EL, Roth, AV, & Fensterseifer, JE 2002, 'Focusing information in manufacturing: a knowledge management perspective', *Industrial Management and Data Systems,* Vol. 102, No. 7, pp.381-389.

Palvia, SC, Sharma, RS, & Conrath, DW 2001, 'A socio-technical framework for quality assessment of computer information systems', *Industrial Management and Data Systems,* Vol. 101, No 5-6, pp. 237-251.

Pande, P, & Holpp, L 2004, *What is Six Sigma?,* McGraw-Hill Education, London.

Pantazi, MA, & Georgopoulos, NB 2006, 'Investigating the impact of business-process-competent information systems (ISs) on business performance', *Managing Service Quality,* Vol. 16, No. 4, pp. 421-434.

Parida, A, & Kumar, U 2006, 'Maintenance performance measurement (MPM): issues and challenges', *Journal of Quality in Maintenance Engineering,* 12(3), pp. 239-251.

Parker, M, Benson, R, & Trainor, HE 1988, *Information Economics: Linking Business Performance to Information Technology,* Prentice-Hall, Englewood Cliffs, NJ.

Parker, MM, Benson, RJ, & Trainor, H E 1997, *Information Economics. Linking Business Performance to Information Technology.* Upper Saddle River, New Jersey.

Patel, NV, Irani, Z 1999, 'Evaluating information technology in dynamic environments: A focus on tailorable information systems', *Logistics Information Management*, Vol. 12, No.1, pp.32-39.

Patterson, KA, Grimm, CM, & Thomas, MC 2003, 'Adopting new technologies for supply chain management', *Transportation Research Part E*, Vol. 39 pp.95-121.

Patton, MQ 1990, *Qualitative Evaluation and Research Methods*, 2nd edition, Sage Publications, Newbury Park, CA.

Pavlou, PA, Housel, TJ, Rodgers, W, & Jansen, E 2005, 'Measuring the return on information technology: A knowledge-based approach for revenue allocation at the process and firm level', *Journal of the AIS*, Vol. 6, No. 7, pp. 199-226.

Pawlowski, SD, & Robey, D 2004, 'Bridging user organizations: Knowledge brokering and the work of information technology professionals', *MIS Quarterly*, Vol. 28, No. 4, pp. 645-672.

Pedler, M, Boydell, T, & Burgoyne, J (1989) 'Towards the learning company', *Management Education and Development*, Vol. 20, No. 1, pp. 1-8.

Pegels, CC, & Watrous, C 2005, 'Application of the theory of constraints to a bottleneck operation in a manufacturing plant', *Journal of Manufacturing Technology Management*, Vol. 16 No. 3, pp. 302-311.

Pennington, D & Wheeler, F 1998 'The Role of Governance in IT Projects: Integrating the Management of IT Benefits', in Proceedings of the Fifth European Conference on IT Investment Evaluation. pp.25-34.

Penrose, E 1995, *The Theory of the Growth of the Firm,* Oxford University Press, NY.

Percy, DF, & Kobbacy, KAH 1996, 'Preventive maintenance modelling - A Bayesian perspective', *Journal of Quality in Maintenance Engineering*, 2(1), pp. 15-24.

Perry, C. 1998a, 'Processes of a case study methodology for postgraduate research in marketing', *European Journal of Marketing*, Vol. 32, No. 9/10, pp. 785-802.

Perry, C, 1998b, A structured approach to presenting theses, *Australian Marketing Journal*, Vol. 6, No. 1, pp. 63-86.

Pollock, TG, Whitbred, RC, & Contractor, N 2000, 'Social information processing and job characteristics: A simultaneous test of two theories with implications for job satisfaction', *Human Communication Research*, Vol. 26, No. 2, pp. 292-330.

Porra, J, Hirschiem, R, & Parks, MS 2005, 'The history of Texaco's corporate information technology function: A general systems theoretical interpretation', *MIS Quarterly*, Vol. 29, No. 4, pp.721-746.

Porter, ME 1979, 'How competitive forces shape strategy', *Harvard Business Review*, Vol. 57, No. 2, pp. 137-145.

Porter, ME, & Miller, VE 1985, 'How information gives you competitive advantage', *Harvard Business Review*, Vol. 63 No.4, pp.149-60.

Porter, ME 1980, *Competitive Strategy*, Free Press, New York.

Porter, ME 1996, 'What is strategy?', *Harvard Business Review*, 74(6), pp. 61-78.

Porter, ME 2001, 'Strategy and the internet', *Harvard Business Review*, 79(3), pp. 63-78.

Pouloudi, A & Whitley, A 1997 'Stakeholder identification in interorganizational systems: gaining insights for drug use management systems', *European Journal of Information Systems*, vol. 6, no. 1, pp.1-14.

Power, D 2005, 'Implementation and use of B2B-enabling technologies: five manufacturing cases', *Journal of Manufacturing Technology Management*, Vol. 16 No. 5, pp. 554-572.

Powell, P 1992, 'Information technology evaluation: Is it different?', *Journal of the Operational Research Society*, Vol. 43, No. 1, pp. 29-42.

Powell, PL 1996, 'Evaluation of Information Technology Investments: Business as Usual?', in *Beyond the IT Productivity Paradox*, eds LP Willcocks, & S Lester, Chichester: John Wiley and Sons, pp. 151-182.

Powell, PL 1999, 'Evaluation of Information Technology Investments: Business as Usual?', in *Beyond the IT productivity paradox*, eds. LP Willcocks & S Lester, John Wiley & Sons, West Sussex, UK.

Powell, S 2004, 'The challenges of performance measurement', *Management Decision*, Vol. 42, No. 8, pp. 1017-1023.

Powell, TC & Dent-Micallef, A 1997, 'Information technology as competitive advantage: the role of human, business and technology resources', *Strategic Management Journal*, Vol. 18, No. 5, pp. 375-405.

Powell, WW, and DiMaggio, PJ (eds.). 1992, 'The new institutionalism in organizational analysis', University of Chicago Press, Chicago, USA.

Pozzebon, M, & Pinsonneault, A 2005, 'Global-local negotiations for implementing configurable packages: the power of initial organizational decisions", *Journal of Strategic Information Systems*, Vol. 14, No. 2, pp. 121-145.

Prahalad, CK & Hamel G 1990, 'The core competence of the corporation', *Harvard Business Review*, Vol. 68, No. 3, pp. 79-91.

Premkumar, G, & King, WR 1994, 'Organizational Characteristics and Information Systems Planning: An Empirical Study', *Information Systems Research*, Vol. 5, No. 2, pp. 75-109

Premkumar, G, Ramamurthy, K, & Saunders, CS 2005, 'Information processing view of organizations: An exploratory examination of fit in the context of interorganizational relationships', *Journal of Management Information Systems*, 22(1), pp. 257-294.

Prickett, PW 1999, 'An integrated approach to autonomous maintenance management', *Integrated Manufacturing Systems*, Vol. 10, No. 4, pp. 233-243.

Pun, KF 2005, 'An empirical investigation of strategy determinants and choices in manufacturing enterprises', *Journal of Manufacturing Technology Management*, Vol. 16, No. 3, pp. 282-301.

Qu, Z, & Brocklehurst, M 2003, 'What will it take for china to become a competitive force in offshore outsourcing? An analysis of the role of transaction costs in supplier selection', *Journal of Information Technology*, Vol. 18, No. 1, pp. 53-67.

Quinn, J, & Baily, M 1994, 'Information technology: The key to service performance', *Brookings Review*, Vol. 12, pp. 36-41.

Quinn, JB, Anderson, P, & Finkelstein, S 1996, 'Managing professional intellect: making the most of the best', *Harvard Business Review*, Vol. 74, No.2, pp.71-80.

Rafaeli, S, & Raban, DR 2003, 'Experimental Investigation of the Subjective Value of Information in Trading', *Journal of the Association for Information Systems*, Vol. 4, pp. 119-139.

Ragowsky, A, & Stern, M 1997, 'The benefits of IS for CIM applications: a survey', *International Journal of Computer Integrated Manufacturing*, 10(1-4), pp.245–255.

Rahm E, & Do HH 2000, 'Data Cleaning: Problems and Current Approaches', *IEEE Bulletin of the Technical Committee on Data Engineering*, Vol. 24, No. 4, pp. 3-13.

Rao, HA, & Gu, P 1997, 'Design methodology and integrated approach for design of manufacturing systems', *Integrated Manufacturing Systems*, 8(3), pp. 159-172.

Ray, CM, Harris, TM & Dye, JL 1994, 'Small business attitudes toward computers', *Journal of End User Computing*, Vol. 6, No. 1, pp. 16-25.

Raymond, L 2005, 'Operations management and advanced manufacturing technologies in SMEs: A contingency approach', *Journal of Manufacturing Technology Management*, Vol. 16 No. 8, pp. 936-955.

Redman, TC 1996, *Data Quality for the Information Age*, Artech House, Norwood MA.

Remenyi, D 1991, *A Guide to Measuring and Managing IT Benefits*, Blackwell Publishers, Inc. Cambridge, MA, USA .

Remenyi, D, & Whittaker, L 1996, 'The evaluation of business process re-engineering projects, In *Investing in information systems: evaluation and management*, Chapter 7, ed. L Willcocks, Chapman & Hall, UK.

Remenyi, D, Williams, B, Money, A, & Swartz, E 1998, *Doing Research in Business and Management*, Sage Publications, London.

Remenyi, D, & Sherwood-Smith, M 1999, 'Maximise information systems value by continuous participative evaluation', *Logistics Information Management*, Vol. 12. No. 1/2, pp. 14-31.

Remenyi, D, Money, A, & Sherwood-Smith, M 2000, The effective measurement and management of IT costs and benefits, 2nd edition, Butterworth Heinemann, Oxford

Robie, C, Tuzinski, KA, & Bly, PR 2006, 'A survey of assessor beliefs and practices related to faking', *Journal of Managerial Psychology*, Vol. 21, No. 7, pp. 669-681.

Robson, C 2004, *Real World Research*, 2nd edition, Blackwell Publishing, Oxford.

Romano, C 1989, 'Research strategies for small business: a case study', *International Small Business Journal*, Vol. 7, No.4, pp.35-43.

Rogers, EM 2003, Diffusion of Innovations, 5th ed. Free Press, New York, USA.

Rondeau, EP, Brown, RK, & Lapides, PD 2006, *Facility Management*, John Wiley & Sons, Hoboken, New Jersey.

Rose, J 2002, 'Interaction, transformation and information systems development – an extended application of soft systems methodology', *Information Technology and People*, Vol. 15, No. 3, pp. 242-268.

Rothenberg, S, Pil, FK, & Maxwell, J 2001, 'Lean, green and the quest for superior performance', *Production and Operations Management*, Vol. 10, No.3, pp.228-243.

RTA 2000, 'RTA:2000 - Roads and Traffic Authority Annual Report 2000', Roads and Traffic Authority, News South Wales Government, Sydney, Australia.

Rudberg, M 2002, Manufacturing strategy: linking competitive priorities, decision categories and manufacturing networks, PROFIL 17, Linkoping Institute of Technology, Linkoping, Sweden.

Ryan, SD, Harrison, DA, & Schkade, LL 2002, 'Information-technology investment decisions: When do costs and benefits in the social subsystem matter?', *Journal of Management Information Systems*, Vol. 19, No. 2, pp. 85-127.

Sabherwal, R 1999, 'The relationship between information system planning sophistication and information system success: an empirical assessment', *Decision Sciences*, Vol. 30 No.1, pp.137-67.

Sabherwal, R, Hirschheim, R, & Goles, T 2001, 'The dynamics of alignment: Insights from a punctuated equilibrium model', *Organization Science*, 12(2), pp. 79-197.

Sahay, S 1997, 'Implementation of information technology: A time-space perspective', *Organization Studies*, Vol. 18, No. 2, pp. 229-260.

Sakaguchi, T, & Dibrell, CC 1998, 'Measurement of the intensity of global information technology usage: quantitizing the value of a firm's information technology', *Industrial Management & Data Systems*, Vol. 98, No. 8, pp. 380-394.

Sakaguchi, T, Nicovich, SG, & Dibrell, CC 2004, 'Empirical evaluation of an integrated supply chain model for small and medium sized firms', *Information Resources Management Journal*, Vol. 17, No. 3, pp. 1-9.

Saleh, Y, & Alshawi, M 2005, 'An alternative model for measuring the success of IS projects: the GPIS model', *The Journal of Enterprise Information Management*, Vol. 18, No. 1, pp. 47-63.

Sambamurthy, V, Bharadwaj, A, & Grover, V 2003, 'Shaping firm agility through digital options: Reconceptualizing the role of it in contemporary firms, *MIS Quarterly*, Vol. 27, No. 2, pp. 237-263.

Sandberg, U 1994, 'The Coupling Between Process and Product Quality – The interplay Between Maintenance and Quality in Manufacturing', in Proceedings of 12th Euromaintenance Conference, European Foundation of National Maintenance Societies, Amsterdam.

Santhanam, R, & Hartono, E 2003, 'Issues in linking information technology capability to firm performance', *MIS Quarterly*, Vol. 27, No. 1, pp 125-153.

Santori, P, & Anderson, AD 1987, 'Manufacturing performance in the 1990s: measuring for excellence', *Journal of Accountancy*, Vol. 164, No. 5, pp. 141-147.

Sardar, G, Ramachandran N, & Gopinath, R 2006, 'Challenges in Achieving Optimal Asset Performance Based on Total Cost of Ownership', in Proceedings of Proceedings of Inaugural World Congress on Engineering Asset Management, eds J. Mathew, L Ma, A Tan & D Anderson, 11-14 July 2006, Gold Coast, Australia

Sarkis, J, & Sundarraj, RP 2000, 'Factors for strategic evaluation of enterprise information technologies', International Journal of Physical Distribution & Logistics Management, Vol. 30, No. 3/4, pp. 196-220.

Sarmento, A 2005, 'Knowledge management: at a cross-way of perspectives and approaches', *Information Resources Management Journal*, Vol. 18 No.1, pp.1-7.

Sauer, C, Yetton, PW (eds) 1997, *Steps to the Future: Fresh Thinking on the Management of IT-Based Organizational Transformation*, Jossey-Bass Publishers, San Francisco.

Saunders, CS, & Jones, JW 1992, 'Measuring performance of the information systems function', *Journal of Management Information Systems*, Vol. 8, No. 4, pp. 63-82.

Schienstock, G 1999, 'Information society, work and the generation of new forms of social exclusion', (SOWING): First Interim Report (Literature Review), Tampere, Finland, accessed online on June 30, 2006, at http://www.uta.fi/laitokset/tyoelama/sowing/frontpage.html

Schilling, MA, Vidal, P, Ployhart, RE, & Marangoni, A 2003, 'Learning by doing something else: Variation, Relatedness, and the Learning Curve', *Management Science*, Vol. 49, No. 1, pp. 39-56.

Schmidt, H, Hantschel, I 2000, 'Verbreitung von Evaluation und Evaluationsforschung in der Wirtschaftsinformatik', in Evaluation und Evaluationsforschung in der Wirtschaftsinformatik', eds LJ Heinrich, & I Hantschel, Munchen, Vienna.

Schmitz, J, Platts, KW 2004, 'Supplier Logistics Performance Measurement: Indications From A Study In The Automotive Industry', *International Journal Of Production Economics*, Vol. 89, pp. 231-243.

Schroeder, RG, Scudder, GD, & Elm, DR 1989, 'Innovation in manufacturing', *Journal of Operations Management*, Vol. 8, No.1, pp.1-15.

Schuman, CA, & Brent, AC 2006, 'Asset life cycle management: towards improving physical asset performance in the process industry', *International Journal of Operations and Production Management*, Vol. 25, No. 6, pp. 566-579.

Schwartzman, HB 1993, *Ethnography in Organisations*, Sage, Newbury Park, CA.

Scott Morton, MS (ed) 1991, 'The Corporation of the 1990s: Information Technology and Organizational Transformation', Oxford University Press.

Scott, J 2000, *Social Network Analysis: A handbook*, 2nd edn, Sage Publications, London.

Scott, SV & Wagner EL 2003, 'Networks, negotiations and new times: The implementation of enterprise resource planning into an academic administration', *Information and Organization*, Vol. 13, Issue 4, pp. 285-313.

Scott, WR 2001, Institutions and organizations, 2nd edition, Thousand Oaks, Sage, CA.

Senge, PM 1990, *The Fifth Discipline: The Age and Practice of the Learning Organization*, Doubleday, New York.

Senge, PM 1996, Leading learning organizations, *Executive Excellence*, 13(4), pp. 10-12.

Serafeimidis, V 1996, 'Information Technology Investment Evaluation: Rational, Concepts and Facilitation', in *Information Technology Management and Organizational Innovations*, ed. M Khosrowpur, Idea Group Publishing, Harrisburg.

Serafeimidis, V 1997, 'Interpreting the Evaluation of Information Systems Investments: Conceptual and Operational Explorations', PhD Thesis, Department of Information Systems London School of Economics and Political Science, London.

Serafeimidis, V, & Smithson, S 2000 'Information Systems Evaluation in Practice: a case study of organisational change', *Journal of Information Technology*, vol. 15, no. 2, pp. 93-105.

Serafeimidis, V, & Smithson, S 2003, 'Information systems evaluation as an organizational institution - experience from a case study', *Information Systems Journal*, vol. 13, no. 3, pp.251-274.

Sethi, A, & Sethi, S 1990, 'Flexibility in manufacturing: a survey', *The International Journal of Flexible Manufacturing Systems*, Vol. 2, pp.289-328.

Shaffner, G 1994, The ultimate test of technology's value, *Computerworld*, 28(37),pp. 37.

Shaft, TM, & Vessey, I 2006, 'The role of cognitive fit in the relationship between software comprehension and modification', *MIS Quarterly*, 30(1), pp. 29-55.

Sharif,AM, & Irani, Z 2006, 'Applying a fuzzy-morphological approach to complexity within management decision making', *Management Decision*, 44(7), pp. 930-961.

Sherman, E, & Reid, WJ 1994, Coming of Age in Social Work - The Emergence of Qualitative Research Sherman, Edmund A. Reid, William James, Qualitative research in social work, New York: Columbia University Press.

Sherwin, D 2000, 'A review of overall models for maintenance management', *Journal of Quality in Maintenance Engineering*, Vol. 6, No. 3, pp. 138-164.

Sherwin, D, & Al-Najjar, B 1999, 'Practical models for condition monitoring inspection intervals', *Journal of Quality in Maintenance Engineering*, 5(3), pp. 203-221.

Shi, N, & Bennett, D 2000, 'Information systems management positions – a market perspective', *Work Study*, Vol. 49, No. 7, pp. 275-284.

Shirose, K, & Goto, F 1989, 'Eliminating the six big losses', in *TPM development program: Implementing total productive maintenance* ed. S. Nakajima, Productivity Press, Cambridge, MA.

Silverman, BS 1999, 'Technological resources and the direction of corporate diversification: toward an integration of the resource-based view and transaction cost economics', *Management Science*, Vol. 45, No. 8, pp. 1109-1124.

Silverstone, R & Haddon, L 1996, 'Design and the Domestication of Information and Communication Technologies: Technical Change and Everyday Life', in *Communication by Design: The Politics of Information and Communication Technologies,* eds R Mansell, & R Silverstone, Oxford University Press, Oxford.

Simmons, P 1996, 'Quality Outcomes: Determining Business Value', *IEEE Software*, vol. 13, no. 1, pp.25-32.

Sink, DS 1991, 'The Role of Measurement in Achieving World Class Quality and Productivity Management', *Industrial Engineering*, Vol. 23, No. 6, pp. 23-30.

Sink, DS, & Tuttle, TC 1989, *Planning and Measurement in Your Organization of the Future*, Industrial Engineering and Management Press, Norcross, GA.

Sink, DS, 1985, 'Productivity Management: Planning, Measurement and Evaluation, Control and Improvement', John Wiley and Sons Inc. New York, NY.

Small, MH, & Chen, IJ 1995, 'Investment justification of advanced manufacturing technology; an empirical analysis', *Journal of Engineering & Technology Management*, Vol. 12, pp.27-55.

Small, MH, & Chen, IJ 1997, 'Economic and Strategic Justification of AMT Inferences from Industrial Practices', *International Journal of Production Economics*, Vol. 49 pp.65-75.

Small, MH, & Yasin, M 2003, 'Advanced manufacturing technology adoption and performance: the role of management information systems departments', *Integrated Manufacturing Systems*, Vol. 14, No. 5, pp. 409-422.

Small, MH 2006, 'Justifying investment in advanced manufacturing technology: a portfolio analysis', *Industrial Management and Data Systems*, 106(4), pp. 485-508.

Smith, AM 1993, *Reliability-centred Maintenance*, McGraw-Hill, New York, NY.

Smith, C, Knezevic, J 1996, 'Achieving quality through supportability - part I: concepts and principles', *Journal of Quality in Maintenance Engineering*, 2(2), pp.3-20.

Smithson, S & Hirschheim, R 1998, 'Analysing information system evaluation: another look at an old problem', *European Journal of Information Systems*, 7(3), pp.158-174.

Song, S 2001, 'An internet knowledge sharing system', *The Journal of Computer Information Systems*, Vol. 42 No.3, pp.25-30.

Songer, AD, Young, R, & Davis, K 2001, 'Social architecture for sustainable IT implementation in AEC/EPC', in Proceedings of IT in Construction in Africa 2001, eds. G Coetzee & F Boshoff, Mpumalunga, 30 May – 1 June, South Africa.

Stake, R 1980, 'The case method inquiry in social inquiry', in *Towards a Science of the Singular*, ed. H Simons, Centre for Applied Research in Education, University of East Anglia, Norwich, CARE Occasional Publications.

Stake, R 1995, *The art of case research,* Sage Publications, Thousand Oaks, CA.

Stalk, G, Evans P, & Shulman LE 1992, 'Competing on capabilities: the new rules of corporate strategy', *Harvard Business Review,* Vol. 70, No. 2, pp. 57-69.

Standing, C, Guilfoyle, A, Lin, C, & Love, PED 2006, 'The attribution of success and failure in IT projects, *Industrial Management & Data Systems,* Vol. 106, No. 8, pp. 1148-1165.

Stapelberg, RF 2006, 'Professional Skills Training in Integrated Asset Management: How to Develop and Implement the Essential Organisational Asset Management Functions', World Congress on Engineering Asset Management, 11-14 July, Gold Coast, Australia

Steadman, M, Albright, T, & Dunn, K 1996, 'Stakeholder group interest in the new manufacturing environment', *Managerial Auditing Journal,* Vol. 11, No. 2, pp. 4-9.

Steenstrup, K, (2007), EAM and IT Enabled Assets: What Is Your Equipment Thinking About Today, unpublished paper.

Stephenson, P, & Blaza, S 2001, 'Implementing technological change in construction organisations', in Proceedings of IT in Construction in Africa 2001, eds. G Coetzee & F Boshoff, Mpumalunga, 30 May – 1 June, South Africa.

Stewart, R, & Mohamed, S 2002, 'Barriers to implementation information technology in developing countries', in Proceedings of the Council for Research and Innovation in Building and Construction-Working group 107 1st International Conference Creating a sustainable construction industry in developing countries, Spier, Stellenbosch, South Africa, 11-13 November.

Stewart, RA, Mohamed, S, & Marosszeky, M 2004, 'An empirical investigation into the link between information technology implementation barriers and coping strategies in the Australian construction industry', *Construction Innovation,* 4(3), pp. 155-171.

Stewart,R, & Mohamed, S 2002, 'IT/IS projects selection using multi-criteria utility theory', *Logistics Information Management,* Vol. 15, No. 4, pp. 254-270.

Street, CT, & Meister, DB 2004, 'Small business growth and internal transparency: The role of information systems', *MIS Quarterly,* Vol. 28, No. 3, pp. 473-506.

Stockdale, R & Standing, C 2006, 'An interpretive approach to evaluating information systems: A content, context, process framework', *European Journal of Operational Research,* vol. 173, no. 3, pp. 1090-1102

Strong, DM 1997, 'IT process designs for improving information quality and reducing exception handling: A simulation experiment', *Information and Management,* Vol. 31, pp. 251-63.

Sudweeks, F, Mclaughlin, ML, & Rafaeli, S (eds.) 1998, *Network and Netplay,* MIT Press, Cambridge MA.

Sun, H, & Riis, JO 1994, 'Organizational, technical, strategic, and managerial issues along the implementation process of advanced manufacturing technology – a general framework of implementation guide', *The International Journal of Human Factors in Manufacturing,* Vol. 4, No.1, pp.23-36.

Sutcliffe, AG 2000, 'Requirements analysis for socio-technical system design', *Information Systems,* Vol. 25, No. 3, pp. 213-233.

Suwardy, T, Ratnatunga, J, Sohal, AS, & Speight, G 2003, 'IT projects: evaluation, outcomes and impediments', Benchmarking: An International Journal, Vol. 10, No. 4, pp. 325-342.

Suwignjo, P, Bititci, US, & Carrie, AS 2000, 'Quantitative models for performance measurement systems', *International Journal of Production Economics*, Vol. 64, pp.231-241.

Sveiby, KE 1997, 'The intangible asset monitor', *Journal of Human Resource Costing andAccounting*, vol. 2, no. 1, pp.73-97.

Tan, DS 1995, 'IT management plateaus: an organizational architecture for IS', *Information Systems Management*, Winter, pp. 44-53.

Tan, DS 1999, 'Stages in information systems management', *Handbook of IS Management*, CRC Press LLC, Boca Raton, FL, pp.51-75.

Tangen, S 2004, 'Performance measurement: from philosophy to practice', *International Journal of Productivity and Performance Management*, Vol. 53 No. 8, pp. 726-737.

Tangen, S 2005, 'Insights from practice Analysing the requirements of performance measurement systems', *Measuring Business Excellence*, Vol. 9, No. 4, pp. 46-54.

Tapinos, E, Dyson, RG, & Meadows, M 2005, 'The impact of performance measurement in strategic planning', *International Journal of Productivity and Performance Management*, Vol. 54 No. 5/6, pp. 370-384.

Tapscott, D, & Caston, A 1993, *Paradigm Shift: The New Promise of Information Technology*, McGraw-Hill, New York.

Tashakkori, A, & Teddlie, C 1998, *Mixed Methodology: Combining Qualitative and Quantitative Approaches,* Sage Publications, London, U.K.

Taskinen, T, & Smeds, R 1999, 'Measuring change project management in manufacturing', *International Journal of Operations and Production Management*, Vol. 19, No. 11, pp. 1168 – 1187.

Tayi, GK, & Ballou, D 1998, 'Examining Data Quality', in *Communications of the ACM*, Vol. 41, No. 2, pp. 54-57.

Teigland, R, & Wasko, M 2003, 'Integrating knowledge through information trading: Examining the relationship between boundary spanning communication and individual performance', *Decision Sciences*, Vol. 34, No. 2, pp 261-287

Teng, S, & Ho, S 1996 'Failure mode and effects analysis: An integrated approach for product design and process control', *International Journal of Quality & Reliability Management*, Vol. 13, No. 5 , pp. 8-26.

Teo, TSH, & King, WR 1997, 'Integration between business planning and information systems planning: an evolutionary-contingency perspective', *Journal of Management Information Systems*, Vol. 14 , No.1, pp.185-224.

Teo, TSH, & Ang, JSK 1999, 'Critical success factors in the alignment of IS plans with business plans', *International Journal of Information Management*, 19(2), pp173-85.

Teo, TSH, & Yu, Y 2005, 'Online buying behavior: A transaction cost economics perspective', *Omega*, Vol. 33, No. 5, pp. 451-465.

Teubner, RA 2005, 'The IT21 Checkup for IT Fitness: Experiences and Empirical Evidence from 4 Years of Evaluation Practice', in working papers, European Research Center for Information Systems No. 2., eds. J Becker, K Backhaus, HL Grob, T Hoeren, S Klein, H Kuchen, U. Muller-Funk, UW Thonemann, G Vossen, Munster, ISSN 1614-7448.

Thorpe, D 2003, 'Online remote construction management trials in Queensland department of main roads: a participant's perspective', *Construction Innovation*, Vol. 3, No. 2, pp. 65-79.

Trauth, EM 1997, 'Achieving the Research Goal with Qualitative Methods: Lessons Learned Along the Way', in *Information Systems and Qualitative Research*, eds. AS Lee J Liebenau & JI DeGross, Chapman and Hall, London, pp. 225-245.

Truex, DP, Baskerville, R, & Klein, H 1999, 'Growing Systems in Emergent Organizations', *Communications of the ACM*, vol. 42, no. 8, August, pp.117-123.

Trygg, LD 1991, 'Engineering design – some aspects of product development efficiency', PhD Thesis, Chalmers University of Technology, Gothenburg.

Tsang, AHC 1995, 'Condition-based maintenance: tools and decision making', *Journal of Quality in Maintenance Engineering*, Vol. 1, No.3, pp.3-17.

Tsang, AHC 1999, 'Maintenance performance management in capital intensive organisations', PhD thesis, Department of Manufacturing Engineering, The Hong Kong Polytechnic University, Hong, Kong.

Tsang, AHC 2002, 'Strategic dimensions of maintenance management', *Journal of Quality in Maintenance Engineering*, Vol. 8, No. 1, pp. 7-39.

Tsang, AHC, Jardine, AKS, & Kolodny, H 1999, 'Measuring maintenance performance: a holistic approach', *International Journal of Operations & Production Management*, Vol. 19, No. 7, pp. 691-715.

Twigg, D 2002, 'Managing the design/manufacturing interface across firms', *Integrated Manufacturing Systems*, Vol. 13. No. 4, pp. 212-221.

UMS, 1999, *Integrated Asset Management Assessment*, UMS Group, Australia Pty Ltd.

Van Der Blonk, H 2000, 'Institutionalization and Legitimation of Information Technologies in Local Contexts', in Proceedings of the Information Flows, Local Improvisations and Work Practices, International Federation of Information Processing Working Group 9.4 on Social Implications of Computers in Developing Countries, Cape Town, May 23-26.

Vatn, J 1997, 'Maintenance optimisation from a decision theoretical point of view', *Reliability Engineering & System Safety*, No. 58, pp. 119-126.

Vaughan, D 1992, 'Theory elaboration: the heuristics of case analysis', in *What is a Case? Exploring the Foundations of Social Inquiry*, eds. CC Ragin HS Becker,Cambridge University Press, Cambridge, pp.173-292.

Venkatesh, V, & Davis, FD 2000, 'A theoretical extension of the technology acceptance model: Four longitudinal field studies', *Management Science*, 46(2), pp. 186-204.

Venkatesh, V, Morris, MG, Davis, GB, & Davis, FD 2003, 'User acceptance of information technology: Toward a unified view', *MIS Quarterly*, 27(3), pp. 425-478.

Verweire, K, & Berghe, LV 2003, Integrated performance management: adding a new dimension, *Management Decision*, Vol. 41, No. 8, pp. 782-790.

Vessey, I 1991, 'Cognitive fit: A theory-based analysis of the graphs versus tables literature', *Decision Sciences*, Vol. 22, No. 2, pp. 219-240.

Vessey, I 2006, 'The theory of cognitive fit: One aspect of a general theory of problem solving?', in *Human-computer interaction and management information systems: Foundations (Advances in Management Information Systems Series)*, eds. P Zhang & D Galletta, M.E. Sharpe Inc., Armonk, NY.

Vessey, I, & Glass, RL 1994, 'Applications-based methodologies', *Information Systems Management*, Vol. 11, No. 4, pp. 53-57.

Victorian Department of Infrastructure 2000, 'Facing the renewal challenge; Victorian Local Government infrastructure study', Melbourne, Victoria, Australia.

Volkow, N 2003, Interaction between information systems and organizational change: case study of Petroleos Mexicanos, PhD Thesis, Department of Information Systems London School of Economics and Political Science, London.

Voordijk, H, Leuven, AV, & Laan, A 2003, 'Enterprise resource planning in a large construction firm: implementation analysis', *Construction Management and Economics*. Vol. 21, No. 5, pp. 511-521.

Wade, M, & Hulland, J 2004, 'The resource-based view and information systems research: Review, extension and suggestions for future research', *MIS Quarterly*, Vol. 28, No. 1, pp. 107-138.

Waeyenbergh, G, & Pintelon, L 2002, '*A framework for maintenance concept development*', *International Journal of Production Economics*, 77(3), pp. 299-313.

Wainwright, D, Green, G, Mitchell, E, & Yarrow, D 2005, 'Towards a framework for benchmarking ICT practice, competence and performance in small firms, *Performance Measurement and Metrics: The International Journal for Library and Information Services*, Vol. 6, No. 1, pp. 39-52.

Walsh, JP, & Ungson, GR 1991, 'Organizational Memory', *Academy of Management Review*, Vol. 16, No. 1, pp. 57-91.

Walsham, G & Sahay, S 1999, 'GIS for district-level administration in India: problems and opportunities', *MIS Quarterly*, Vol. 23, No. 1, pp. 39-65.

Walsham, G 1993, *Interpreting Information Systems Research in Organizations*, John Wiley, Chichester.

Walsham, G 1995, 'Interpretive Case Studies in IS Research: Nature and Method', *European Journal of Information Systems*. Vol.4, No.2, pp. 74-83.

Walsham, G 2001, *Making a World of Difference IT in a Global Context*, John Wiley, Chichester.

Walsham, G 2002, 'Cross-cultural software production and use: A structurational analysis', *MIS Quarterly*, Vol. 26, No. 4, pp. 359-380.

Walter, SG, & Spitta, T 2004, Approaches to the Ex-ante Evaluation of Investments into Information Systems', *WIRTSCHFTSINFORMATIK, WI – State-of-the-Art*, Vol. 46, No. 3, pp. 171-180.

Walther, JB 1995, 'Relational aspects of computer-mediated communication', *Organization Science*, Vol. 6, No. 2, pp. 186-203.

Wang, C, & Ahmed, P 2003, 'Organisational learning: a critical review', *The Learning Organisation*, Vol. 10, No. 1, pp. 8-17.

Wang, RY, Strong, D, & Guarascio, LM 1994, 'Beyond Accuracy: What Data Quality Means to Data Consumers', Technical Report TDQM-94-10, Total Data Quality Management Research Program, Sloan School of Management, Cambridge, MA.

Wang, RY, Storey, VC, & Firth, CP 1995, 'A Framework for Analysis of Data Quality Research', *IEEE Transactions on Knowledge and Data Engineering*, Vol. 7, No. 4, pp. 623-640.

Wang, RY, & Strong, DM 1996, 'Beyond Accuracy: What Data Quality Means to Data Consumers', *Journal of Management Information Systems*, Vol. 12 No. 4, pp. 5-33.

Wang, S 1999, 'An object-oriented approach to plant configuration management information systems analysis', *Industrial Management & Data Systems*, Vol. 99. No. 4, pp. 159-167.

Ward, J, Taylor, P, & Bond, P 1996, 'Evaluation and realization of IS/IT benefits: An empirical study of current practice', *European Journal of Information Systems*, Vol. 4, pp. 214–225.

Ward, P, Duray, GK, Leong, GK, & Sum, CC 1995, 'Business environment, operations strategy, and performance: an empirical study of Singapore manufacturers', *Journal of Operations Management*, Vol. 13 No.2, pp.99-115.

Weick, KE 1995, *Sensemaking in Organizations*, Sage Publications, Beverly Hills, CA.

Weill, P, Subramani, M, & Broadbent, M 2002, 'Building IT infrastructure for strategic agility', *Sloan Management Review*, Vol. 44, No. 1, pp. 57-65.

Weippert, A, Kajewski, SL, & Tilley, PA 2002, 'Internet-based information and communication systems on remote construction projects: A case study analysis', *Construction Innovation*, Vol. 2, No. 2, pp. 103-116.

Wernerfelt, B 1984, 'A resource-based view of the firm', *Strategic Management Journal*, Vol. 5, pp.171-180.

Whiting, R, Davies, J, & Knul, M 1996, Investment Appraisal for IT systems, in *Investing in Information Systems*, ed. L Willcocks, Chapman and Hall, London.

Whitworth, B, & De Moor, A 2003, 'Legitimate by design: Towards trusted socio-technical systems', *Behaviour and Information Technology*, 22(1), pp. 31-51.

Whyte, J, & Bouchlaghem, D 2001, 'IT Innovation within the Construction Organisation', in Proceedings of IT in Construction in Africa 2001, eds. G Coetzee & F Boshoff, Mpumalunga, 30 May-1 June, South Africa.

Wieder, B, Booth, P, Matolcsy, ZP, & Ossimitz, M 2006, The impact of ERP systems on firm and business process performance, *Journal of Enterprise Information Management*, Vol 19, No. 1, pp. 13-29.

Willcocks, L 1996, 'Introduction: Beyond the IT Productivity Paradox', in *Investing in Information Systems: Evaluation and Management*', ed. L. Willcocks, Chapman and Hall, London.

Willcocks, LP, & Lester, S 1997, 'In search of information technology productivity: Assessment issues', *Journal of Operations Research Society*, vol. 48 , pp1082-1094.

Wixon, D, & Ramey, J (eds.) 1996, *Field methods casebook for software design*, John Wiley, New York.

Womack, JP, & Jones, DT 2003, *Lean Thinking: Banish waste and create wealth in your corporation*, Free Press, New York, NY.

Wongrassamee, S, Gardiner, PD, & Simmons, JEL 2003, 'Performance measurement tools: the balanced scorecard and the EFQM excellence model, *Measuring Business Excellence*, Vol. 7, No. 1, pp. 14-29.

Woodhouse, J 2001, 'Asset Management', The Woodhouse Partnership Ltd, accessed on September 10, 2005 at http://www.plantmaintenance.comlarticles/AMbasicintro.pdf

Woodward, DG 1997, 'Life cycle costing – theory, information acquisition and application', *International Journal of Project Management*, 15(6), pp. 335-44.

Xu, H 2003, *Critical Success Factors for Accounting Information Systems Data Quality*, PhD Thesis, University of Southern Queensland, Towoomba, Australia.

Wu, B 2001, 'A unified framework of manufacturing systems design', *Industrial Management & Data Systems*, Vol. 101, No. 9, pp. 446-469.

Yamashina, H 2000, 'Challenge to world class manufacturing', *International Journal of Quality & Reliability Management*, Vol. 17, No. 2, pp. 132-143.

Yin, RK 1994, *Case Study Research, Design and Methods,* 2nd edition, Sage Publications, Newbury Park, CA.

Ying-Pin, Y 2005, 'Identification of factors affecting continuity of cooperative electronic supply chain relationships: Empirical case of the Taiwanese motor industry', *Supply Chain Management: An International Journal,* Vol. 10, No. 4, pp. 327-335.

Ylipaa, T 2000, 'High-Reliability Manufacturing Systems', Licentiate Thesis, Department of Human Factors Engineering, Chalmers University of Technology, Goteborg, Sweden.

Yoh, E, Damhorst, ML, Sapp, S, & Laczniak, R 2003, Consumer adoption of the internet: The case of apparel shopping, *Psychology & Marketing,* 20(12), pp1095-1118.

Yu, C 2005, 'Causes influencing the effectiveness of the post-implementation ERP system', *Industrial Management & Data Systems,* Vol. 105, No. 1, pp. 115-132.

Yurdakul, M 2002, 'Measuring a manufacturing system's performance using Saaty's system with feedback approach', *Integrated Manufacturing Systems,* 13(1), pp25-35.

Yusof, MM, Paul, RJ, & Stergioulas, LK 2006, 'Towards a Framework for Health Information Systems Evaluation', in Proceedings of the 39th Annual Hawaii International Conference on System Sciences, *HICSS '06,* vol. 5, Hawaii.

Zacharia, ZG, & Mentzer, JT 2004, 'Logistics salience in a changing environment', *Journal of Business Logistics,* Vol. 25, No. 1, pp. 187-210.

Zaheer, A & Dirks, K 1999, 'Research on strategic information technology: A resource-based perspective', in *Strategic Management and Information Technology,* eds. N Venkatraman, JC Henderson, JAI Press, Greenwich, CT.

Zahra, SA & George, G 2002, 'The net-enabled business innovation cycle and the evolution of dynamic capabilities, *Information Systems Research,* 13(2), pp.147-150.

Zakaria, N, Stanton, JM, & Sarkar-Barney, STM 2003, 'Designing and implementing culturally-sensitive IT applications - The interaction of culture values and privacy issues in the Middle East, *Information Technology & People,* 16(1), pp. 49-75.

Ziaee, M, Fathian, M, & Sadjadi, SJ 2006, 'A modular approach to ERP system selection - A case study', *Information Management & Computer Security,* 14(5), pp. 485-495.

Zipf, PJ 2000, 'Technology-Enhanced Project Management', *Journal of Management in Engineering,* Vol. 16, Issue 1, pp. 34-39.

Zmud, RVJ 1988, 'Building Relationships Throughout the Corporate Entity', Chapter 3 in Transforming the IS Organization, ed. JJ Elamet al, ICIT Press, Washington DC.

Zmud, RW 1982, 'Diffusion of modern software practices: Influence of centralization and formalization. *Management Science,* Vol. 28, No. 12, pp. 1421-1431.

Zuboff, S 1988, *In the age of the smart machine,* Basic Books, New York.

Appendix A

Barriers to IS Implementation in Engineering Organisations

(Literature Review 2000 -2007)

Reference	Scope	Barriers		
		Operational Level	*Management Level*	*Strategic Level*
Marsh and Flanagan (2000)	Study of drivers and barriers of technology adoption among different industry sectors, primarily manufacturing.	Variety of disparate IT/OT platforms; ineffective application integration and information interoperability; ignorance of importance of data quality.	Adhoc planning leading to improvised IT solutions; employee resistance to change; inability to justify investments in IT adoption.	Technological conservatism; short term business relationships hampering maturity of technology.
Marosszeky et al. (2000)	Study in Australian construction industry identifying levels of IT implementation and risk factors.	Fragmented approach to technology implementation.	Low level of trust among business partners.	Narrow scope and limited vision of strategic use of IT; IT investment decisions driven by cost considerations.
O'Brien (2000)	Study of issue relating to e-commerce technology implementation in engineering enterprises.	Lack of fit of technology with the business processes; information access and usage restrictions; ill-defined information exchange structure.	Lack of job redesign as a result of technology adoption; expectations from technology outclassing technical capability of the organisation.	Legal and cost barriers.
Zipf (2000)	Study of project based e-commerce technologies in an engineering organisation.	Lack of up-skilling and training on new technology.	Inability to allocate financial and non financial resources to support technology implementation; evaluation of effectiveness of IT solutions.	Lack of technology acceptance and change management; lack of management commitment; lack of technology need assessment.
Abdel-Malek et al. (2000)	Study of technology implementation in manufacturing organisations.	Inability to maintain quality of information; skills and people attitude towards technology; technology acceptance and change; lack of technology integration.	Mismatch of technical solution with organizational infrastructure; lack of process control; lack of involvement of various organisational levels in technology adoption process.	Top management not convinced of economic benefits and likelihood that these will be realized; inability to assess future requirements and information needs.
Whyte and Bouchlaghem (2001)	Study of issues in virtual reality applications implementation among design managers in Africa.	Lack of data standards and systems support; slowness of technology; unexpected technical issues and problems; differences in actual performance and capabilities offered by the off the shelf applications.	Lack of resources to support technology implementation; inability to coordinate technical and business staff.	Lack of wider organisational representation in decision making for investment in technology.
Songer et al. (2001)	Study of individuals from 34 engineering organisations in US focusing on social barriers to technology implementation, for technologies relating to 3D design and simulation, data warehouse, engineering applications, and information management.	Incompatibility of OT; lack of IT/OT integration; lack of supportive organisational culture impeding employees to share knowledge; lack of employee motivation to up-skill.	Lack of awareness of the importance of information management; non-cooperative corporate culture.	High costs of implementation; invisibility of value from IT investment.

Continued next page

Reference	Scope	Barriers		
		Operational Level	*Management Level*	*Strategic Level*
Chan et al. (2001)	Study aimed at providing guidance for manufacturing companies that are preparing to invest in advanced manufacturing technology.	Incompatibility with existing technologies; lack of research and development into what technology suits the business; insufficient level of confidence in certain technologies.	Inappropriate IT evaluation techniques; high attention paid to technical development, but not enough to adjustments needed to accommodate technology; inability to measure soft benefits from IT investments.	IT investment policies primarily driven by financial concerns; lack of awareness of strategic role of technology by management; inconsistent nature of corporate IT/OT governance.
Stephenson and Blaza (2001)	Study focusing on organisational change aimed at successful IT implementation.	Lack of IT and OT compatibility within the organisation to support cross organisation functionality; employee resistance to change; lack of requisite skill base; lack of employee motivation to learn new technologies.	Lack of user involvement in technology adoption process; middle management's resistance to adopt new technology for uncertainties regarding output delivery; lack of organisational fit with technology.	Lack of planning and communication of IT investment rationale to all levels in the organisation; lack of strategic vision; high costs of IT investment and support.
Love et al.(2001)	Study of issues in successful implementation of e-commerce technologies in Australian engineering organisations.	Lack of appropriate IT infrastructure to enable business processes; information security issues; lack of awareness of information quality; lack of skill base, high turnover of employees; resistance to change.	Lack of information exchange between sites; inability or difficulty to measure benefits of IT investments; cost of IT maintenance, and training.	Management's expectations of achieving benefits in the short term; high indirect or hidden costs of IT investment; lack of organisational integration.
Stewart and Mohamed (2002)	Study of barriers to IT implementation in engineering organisations in developing countries.	Lack of quality IT infrastructure; lack of system compatibility; lack of information interoperability; unavailability of skill base.	Lack of awareness of multidisciplinary nature of IT; lack of support from middle managers; high staff workload.	Industrial fragmentation; high cost of IT investments; decreased profit margins.
Weippert et al.(2002)	Study that identifies IT implementation success factors.	Compatibility of technologies; information accessibility; information reliability; quality of information and data input.	Lack of user's involvement in IT adoption choices; lack of training and technical support.	Narrow focus of management in making choices about technology investment.
Bjork (2002)	Study of user attitudes to electronic data management systems.	Slow processing speed; lack of data and data communication standards; employee resistance to change; lack of appropriatee change management stratgeis; varying user attitude towards technology adoption.	Lack of resources for technology support and optimal utilisation.	Organisational functional silos driving technology adoption strategies.
Paiva et al. (2002)	Study of importance of information to knowledge management in manufacturing organisations.	Lack of access to information, information accuracy, timeliness of information.	Mismatch between information needs of the organisation and IS; lack of data sharing between business partners.	Inability of top management to view information as an asset.

Continued next page

Reference	Scope	Barriers		
		Operational Level	*Management Level*	*Strategic Level*
Gordon and Gordon (2002)	Study of Dutch and US based manufacturing organisations IT management.	Lack of requisite hardware and software infrastructure.	Lack of IT coordination and control; non supportive organisational culture and structure.	High degree of IT centralisation, structure, and scope; IT expertise rather than business need driving IT investment decisions.
Alshawi and Ingirige (2003)	Study of benefits and problems of web enabled IT applications in engineering organisations	IT incompatibility; lack of information security infrastructure, skill base and competence to operate technology; inefficient information exchange and communication speed.	Lack of IT support decision making and resource allocation; lack of coordination between project participates	Lack of collaboration between business partners; technology not responding to changing business needs.
Voordijk *et al.* (2003)	Study of success and failure of ERP in Dutch engineering organisations	Lack of fit between IT investments and IT infrastructure maturity.	Inability to match technology implementation methods and change management process.	Lack of fit between business strategy and IT.
Huang *et al.*(2003)	Study of the essential criteria for IT adoption in engineering enterprises	Individual's perception of technology; lack of IT/OT compatibility; inability to keep up with changes in technology.	Low degree of innovativeness in the organisation; hierarchical organisational structure; organisational culture not conducive to IT.	Lack of responsiveness to changes in competitive environment.
Thorpe (2003)	Study of IT implementation issues of online construction management.	Unreliable technology; slow speed of operation; user reluctance to adapt to technology; lacking information security and skill base.	Technology not mature enough to handle information needs of the organisation; benefits of IT utilisation not fully perceived; lack of commitment from technology stakeholders.	High costs of IT investments.
Nitithamyong and Skibniewski (2004)	Study of web-based project management services in engineering organisations.	Lack of information interoperability; technology not mature; resistance to change.	Lack of information ownership; lack of accountability.	Inability to quantify IT investment costs and benefits.
Stewart *et al.*(2004)	Study of barriers to IT implementation at industrial, organisational, and project level in construction industry.	Lacking security and privacy; poor information interoperability; employee resistance to change; lack of skills.	Low levels of awareness of IT benefits; lack of creative culture; inability to measure benefits of IT.	Lack of strategic focus of IT investments; technological conservatism; limited financial resources for IT.
Gomes et al. (2004)	Study of performance measurement literature in manufacturing organisations from 1988 – 2000.	Short term focus on process automation; inability to appreciate the multi-dimensional nature of technology implementation.	Inability to take financial and well as non financial benefits of IT investments in performance evaluation methods; inability to effect change management to adapt to technology; inability to detect implementation issues..	Lack of IT implementation as a means of business strategy translation; lack of matching, organisational objectives, customer needs, and organisational success factors with IT investments.

Continued next page

Reference	Scope	Barriers		
		Operational Level	*Management Level*	*Strategic Level*
Lee (2004)	Study of a business process integrated IT evaluation methodology that integrates business strategy, business process design, and supporting IT investment.	Lack of fit between IT and business processes.	Inability to redesign business processes to adapt to new technology; inability to properly measure process requirements, and manage IT configuration.	Lack of strategic analysis of impact of IT investments.
Abdel-Makoud (2004)	Study of relationship between shop floor technologies and organisational and environmental factors in manufacturing organisation in the UK.	Inability to integrate IT and OT; lack of user involvement in technology implementation process; lack of skills to operate technology.	Non availability of feedback on technology use and its impact of different business areas.	Lack of information on competitive environment.
Pun (2005)	Study of Shanghai and Hong Kong based manufacturing organisations to identify and prioritise the strategy determinants for manufacturing enterprises.	Lack of research and development capabilities into technology investments; lack of employee skills and competencies.	Lack of fit of IT infrastructure with business, objectives.	Inability of technology to contribute to horizontal/vertical integration.
Power (2005)	Study of manufacturing organisations to determine the extent to which long-established technologies (such as electronic data interchange) have been applied across supply chains; factors influencing implementation; and future trends of technologies.	Technology not properly mapped to the process needs.	Lack of understanding of impact of IT; lack of intra organisational collaboration; inability of management to identify and manage IT risks before they become issues.	Lack of top management commitment to institutionalise technology.
Dangayach and Deshmukh (2005)	Study of advanced manufacturing technologies in Indian manufacturing organisations.	Lack of proper requirement analysis and conceptual design of investments in IT; inadequate training	Lack of pre/post implementation evaluation of IT investments.	Inability to view IT investments as source of strategic benefits.
Laurindo and de Carvalho (2005)	Study of manufacturing firms aiming to link enhanced performance of product development processes with the increasing use of IT applications.	IT applications not on par with user demands; lack of application integration; lack of information sharing; non availability of requisite technical support.	Lack of fit between organisational infrastructure, processes, and technology.	Inability to assess impact of IT on strategic orientation; non availability of an IT strategy.
Jaska et al. (2006)	Study aiming at value attributes related to business knowledge and competence of IT personnel within manufacturing organisations.	Ineffective operational support to back IS implementation; passive IT staff; lack of requisite IT skill base.	Lack of quality conscious IT culture; Lack of appropriate IT evaluation techniques.	Lack of organisational responsiveness to make choices as to when and how to migrate to a new technology.
Gindy et al. (2006)	Study of an integrated technology road-mapping methodology for manufacturing organisations that enables management to define its technology requirements and to create a balanced technology project portfolio.	Lack of consensus on technology adoption between different functions.	; lack of integrated approach to IT/OT technology management; inability to identify gaps in technological platforms, prioritisation of technical issues, and creation of action plans, and communication of technology need across the organisation.	Lack of evaluation methodologies for technology acquisition projects, which incorporate organisational, financial, and social factors; inability of IT to provide decision support for business responsiveness and competitiveness.

Continued next page

Reference	Scope	Barriers		
		Operational Level	*Management Level*	*Strategic Level*
Small (2006)	Study aiming at justification of investments in advanced manufacturing technology at manufacturing plants in America.	Lack of consideration of the organizational changes necessitated by technology implementation.	Lack of functional integration.	Inability to evaluate technology before implementation; inability of management to adopt an approach IT implementation that accounts for operational and strategic value of IT.

Appendix B

Theoretical Perspectives on IS Implementation

Theory	Description	Focus	References from IS Literature
Actor Network Theory	Emphasises importance of actors (including organisation, people, and objects such as hardware, software, hardware) to a social network. Order in organisations is maintained through smooth running and interaction of these actors.	Heterogeneous network of social and technical actors to create order.	Callon (1986); Orlikowski *et al.* (1996); McMaster *et al.* (1997); Walsham and Sahay (1999); Larsen *et al.* (1999); Scott and Wagner (2003)
Adaptive Structuration Theory	Based on Giddens' (1984) structuration theory, it states that production and reproduction of the social systems through members' use of rules and resources in interaction.	Structure of IT, organizational environment, and tasks aimed at efficiency.	Hoxmeier *et al.* (2000); Walsham (2002); Barry and Crant (2000); Griffith *et al.* (2003); Hinds and Bailey (2003); Barrett and Scott (2004)
Agency Theory	Study of ubiquitous agency and principle relationships, in which the principal delegates work to an agent. Agency theory takes care of two issues that arise out of such a relationship. Firstly, the conflicts between the aims of the principal and the, secondly, the inability of the principle to verify the behaviour of the agent.	Efficiency through alignment of inter-ests, risk sharing, and contracting	Bahli and Rivard (2003); Mahaney and Lederer (2003); Keil, *et al.* (2004); Kohli and Kettinger (2004); Mirchandani and Lederer (2004); Chen and Edgington (2005); Jasperson *et al.* (2005)
Absorptive Capacity	Emphasises organisations to establish internal R&D capacities, which aid IS development along existing familiarity of technology, as well as through evaluation and incorporation of externally generated technical knowledge.	Capabilities though quantity of knowl-edge absorption	Cohen and Levinthal (1990); Aladwani (2002); Pawlowski and Robey, (2004); Schilling, *et al.* (2003); Malhotra *et al.* (2005); Ko *et al.* (2005)
Cognitive Fit	Developed by Vessey (1991), it proposes that there is a link between information presentation and the tasks enabled by the information. This relationship defines performance of the task for individual users	Problem resolution; process enhance-ment; task perform-ance	Vessey (1991); Vessey and Glass (1994); Beckman (2002); Dunn and Grabski (2001); Mahoney *et al.* (2003); Shaft and Vessey (2006); Vessey (2006)
Critical Social	It suggests that social reality has historical underpinnings and is constituted and reconsti-tuted by the people. Even though people or organisations can mindfully make an effort to alter their social and economic conditions, however their ability to do so is hampered by the forms of social, cultural and political domination. It focuses on the conflicts and contradictions in the social environment and seeks to be a source of emancipation to allevi-ate dissonance.	Learning by doing; social emancipation	Basden (2002); Adam (2002); Dennis and Garfield (2003); Heng and de Moor (2003); Henwood and Hart (2003); Alstyne and Brynjolfsson (2005); Majchrzak *et al.* (2005)
Contingency	Optimal organizational performance is contin-gent upon various internal and external con-straints. Important postulates of this theory are that, a. There is no one best way to manage an organisation, b. 'Fit' between the organisation and its subsystems, c. Successful organisations are able to extend this fit to the organisational envi-ronment, and d. Organisational design and management must satisfy the nature of the task and the work groups.	Organisational Efficiency	Ginberg (1980); Zmud (1982); Barki *et al.* (2001); Becerra-Fernandez and Sabherwal 2001); Chin *et al.* (2003); Khazanchi (2005)
Dynamic capabilities	Stresses integration, building, and reconfigura-tion of organisational competencies (external as well as internal) to address the changing business environment.	Competitiveness	Weill *et al.* (2002); Zahra and George (2002); Daniel and Wilson (2003); Sambamurthy *et al.* (2003); Wade and Hulland (2004);

Continued next page

Theory	Description	Focus	References from IS Literature
Information Processing	It suggests that learning should be approached through the study of memory. It is based on two ideas proposed by Miller (1956). Firstly, the concept of 'chunking and the limited capacity' which posits that the short term memory could hold 5-9 chunks of meaningful information. The second concept of information processing, mimics human capabilities of information processing.	Learning by doing; knowledge reuse	Anandarajan and Arinze (1998); Argyres (1999); Andres and Zmud (2001); Burke et al. (2001); Cooper and Wolfe (2005); Gattiker and Goodhue (2005); Premkumar et al. (2005)
Knowledge based theory of the firm	Treats knowledge as most strategically important resource of an organisation, due mainly to social complexity and difficulty of imitation of knowledge based resources. Organisational knowledge and competencies are therefore chief determinants of enhanced organisational performance and sustained competitive advantage.	Core competencies; sustained competitive advantage	Alavi and Leidner (2001); Huseyin (2005); Teigland and Wasko (2003); Pavlou et al. (2005); Massey and Montoya-Weiss (2006)
Punctuated Equilibrium	In terms of organizational behaviour, this theory comprises three elements, i.e. deep structures, equilibrium periods, and revolutionary periods. Deep structures are the sets of basic choices that a system is composed of, i.e. the fundamental parts into which its units are organised, and the fundamental activity patterns in maintaining the existence of the systems. Equilibrium period is the maintenance of organisational structure and activity patterns with small scale incremental changes made to the systems for it to adapt to the changing environment, without affecting the deep structures. Revolutionary periods occur when deep structures are changed leading to a disorderly state, until choices are made to enact new structures for the system.	Strategic change	Newman and Robey (1992); Loch and Huberman (1999); Sabherwal et al. (2001); Street and Meister (2004); Jarvenpaa et al. (2004); Jasperson et al. (2005); Porra et al. (2005);
Resources Based View	Business organisations possess resources that enable them to gain competitive advantage. Rare resources lead an organisation to sustainable competitive advantage, until the organisation is able to protect against resource imitation, transfer, or substitution.	Competitive advantage	Silverman (1999); Zaheer and Dirks (1999); Gregoire et al. (2001); Hidding (2001); Santhanam and Hartono (2003); Melville et al. (2004)
Resource Dependency	Seeks the organisations to alter their behaviour and structures to acquire and maintain required resources. This also includes modifying their dependence relationships to assume a status of power, that is, by minimizing own dependence on other organisations or by increasing the dependence of other organizations on them.	Organisational dominance	Humphreys et al. (2001); Iskandar et al. (2002); Kern et al. (2002); Zacharia and Mentzer (2004); Sakaguchi et al. (2004); Ying-Pin (2005)
Reason Based Action	It argues that behaviours of individuals are characterised by behavioural intentions, whereas behavioural intentions are themselves derived from the attitude of individuals towards the behaviour and the norms associated with the behavioural performance.	System behaviour	Bobbitt and Dabholkar (2001); Venkatesh et al. (2003); Yoh et al. (2003); Bagchi et al. (2003); Leonard et al. (2004); Hansen et al. (2004); Jae-Nam and Young-Gul (2005)
Systems	Instead of considering a system's properties or their parts or elements, this theory advocates the relationships and understanding of the parts that collectively form the whole, i.e. the system. It includes understanding of system boundaries, input, output, processes, hierarchy, orientation, and flow of information.	System throughput; feedback; control	Churchman (1994); Alter (2001); Sabherwal et al.(2001); Garrity (2002); Markus and Majchrzak (2002); Mora et al.(2003); Chung et al.(2005)

Continued next page

Theory	Description	Focus	References from IS Literature
Social cognitive	It provides a framework for Social cognitive theory provides a framework for understanding, foreseeing, and altering human behaviour. It acknowledges human behaviour as the interaction between, individual traits, actions/behaviour, and environment.	Organisational learning	Compeau et al. (1999); Bolt et al. (2001); Chan and Lu (2004); Hasan and Ali (2004); Kuo et al. (2004); Liaw et al. (2006)
Social Network	It views social as nodes and ties. Nodes represent individual actors in networks, and ties represent the association between them. These relationships can take many forms, however, in its fundamental type a social network represents the relationship between nodes and may be used to investigate social/intellectual capital contained at each node.	Knowledge diffusion; communication strength	Burkhardt (1994); Walther (1995); Feeley and Barnett (1996); Sudweeks et al. (1998); Pollock et al. (2000); Scott (2000); Madey et al. (2002)
Structuration	It attempts to reconcile theoretical duality of social systems such as agency/structure, subjective/objective, and micro/macro perspectives. It does not concentrate on individual entities, but focuses on the social practices ordered across space and time (Giddens 1984). Such a view helps in understanding technology enabled contemporary businesses.	Structure; social system	Sahay (1997); Walsham and Sahay (1999); Newman and Robey (1992); Orlikowski (2000); Orlikowski and Barley (2001); Rose (2002); Jones and Karsten (2003); Pozzebon and Pinsonneault (2005)
Socio Technical	It is build around two organisational sub systems, i.e. technical, which consist of tools and techniques to transform inputs into outputs, and social system, which consists of employees, skills, authority structure, knowledge, behaviours, and values. Socio technical theory is built upon the fit by the collective optimization of these systems. This requires an explicit recognition of the interdependency of these systems.	Process optimization; organisational integration	Sutcliffe (2000); Mumford (2000); Palvia et al. (2001); Ryan et al. (2002); Kling et al. (2003); Lamb and Kling (2003); Whitworth and De Moor (2003)
Soft Systems methodology	It intends to resolve soft and hard issues related to ill structured problems having social impacts. It emphasises that the investigator must taken into account issues other than mere technical. Developed by Checkland (1981), it has seven stages.	Problem resolution	Ledington and Ledington (1999); Jagodzinski et al. (2000); Atkinson (2000); Bausch (2002); Janson and Cecez-Kecmanovic (2005)
Strategic competitiveness	Developed by Porter (1979), it provides roadmap of an organisation's competitiveness through the five forces analysis, value chain analysis, and strategic sets, aimed at providing cost leadership, differentiation, or focused advantages to the organisation.	Competitive forces; competitiveness analysis	Kim and Michelman (1990); Chakravarthy (1997); Chen et al. (2000); Chen et al.(2001); Porter (2001)
Transaction cost economics	It argues that the total costs incurred by an organisation can be divided into two categories, i.e. transaction and production costs. Transaction costs represent all costs that arise from processing of information to organize and synchronize the tasks performed by people and machines to accomplish the primary processes of the organisation. On the other hand, production costs are the costs incurred on producing or creating the good or services through primary processes. It aims to reduce the costs through efficient information processing.	Governance structure; outsourcing, inter organizational coordination and collaboration	Garicano and Kaplan (2001); Bahli and Rivard (2003); Qu and Brocklehurst(2003); Clemons and Hitt (2004); Kauffman and Mohtadi (2004); Teo and Yu (2005); Cannel and Nicholson (2005)
Task Technology Fit	Use of IT is expected to have a positive effect on people's performance, if the capabilities of the technology match the task that people have to perform (Goodhue and Thompson, 1995).	Technology fit; individual performance, system utilization	Goodhue and Thompson, 1995); Lim and Benbasat (2000); Dennis et al. (2001); Gebauer and Shaw (2004)

Appendix C
Performance Evaluation in Engineering Organisations
(Literature Review 2000 -2007)

Authors	Scope	Description	Methodology Employed
Duberley *et al.* (2000)	Manufacturing strategy; manufacturing management; organizational change; organizational culture.	This study summarises research undertaken into nature and behaviour of change brought about by performance evaluation and control systems. This study presents an alternative view and illustrates the ways in which performance evaluation and control systems provide a formative context, which stresses that change requires an understanding of the cultural assumptions underpinning both current and desired systems. Consequently it is important to analyse the values which govern behaviour and uncover the underlying and often unconscious assumptions that influence behaviour.	Qualitative case study involving use of repertory grids, in-depth interviews and observation to examine the impact of performance evaluation control systems on behaviour and nature of change.
Kotha and Swamidass (2000)	Advanced manufacturing technology; business strategy	The study examines the connection between strategy, advanced manufacturing technology and performance measurement using survey methodology in US manufacturing organizations. In contrast to previous studies that emphasize only the flexibility dimension of AMT, this study adopts a multidimensional view and concludes that a fit between strategy and advanced manufacturing technologies leads to superior performance.	Quantitative survey responses from 160 U.S. manufacturing organisations covering four dimensions of manufacturing technology, i.e. information exchange and planning ; product design; low volume flexible automation technology; and high automation.
Mapes *et al.* (2000)	Asset performance; process management	This research study determines the factors that enable an asset to simultaneously achieve high labour productivity, fast, reliable delivery and high quality consistency. The performance evaluation system developed in the study recognises four aspects, (i.e. high adherence to schedule, low process time variability, high reliability of delivery by suppliers, and low variability in process output) and two intermediate measures (i.e. faster throughput times, and lower stock levels), and is centred around four measures of performance, i.e. productivity; quality consistency; customer lead times; and delivery reliability. On the basis of performance it divides assets into three groups, i.e. high performers, medium performers and low performers.	Statistical analysis of performance measure i.e. productivity; quality consistency; customer lead times; and delivery reliability, to calculate a composite performance measure for each of 953 manufacturing assets.
Golden and Powell (2000)	Organizational flexibility; metrics; information technology evaluation.	Investigates the role of IT as a tool to provide flexibility aimed at responsiveness to internal and external changes, process efficiency, and strategic renewal. It proposes that extent of flexibility, which is based on four dimensions, i.e. temporal, range, intention and focus, can be measured by its metrics; efficiency, responsiveness, versatility and robustness. It suggests that these four metrics measure the temporal and range dimensions, and the intention and focus dimensions are operationalised within the context of the specific IT to be evaluated.	Quantative metrics based approach.

Continued next page

Authors	Scope	Description	Methodology Employed
Kutucuoglu et al. (2001)	Asset maintenance, maintenance efficiency; maintenance tasks efficiency	It argues employee involvement, cross functional collaboration, and integration of objective and subjective measures of evaluation with strategic objectives of the organisation. In so doing, it suggests selection of performance indicators in relations to strategic objectives of the organisation, such that each measure provides feedback to these objectives. The PES thus developed targets vertical alignment, and evaluates strategic objectives at operational and strategic levels.	Ex post evaluation of performance using scoring approach using a PES based on quality function deployment (QFD).
De Toni and Tonchia (2001)	Manufacturing strategy; production costs; productivity; quality; flexibility; inter organisational integration	The study reports the results of a survey of 115 medium and large sized manufacturing companies operating in mechanical, electric, and electromechanical industries and concludes that majority of PES are of 'frustum', where traditional cost performances (the production costs and the productivity) are kept separate from the more innovative non-cost measures (quality, time and flexibility). With regard to characteristic of a PES, the study argues formalised relationship between objects and the objectives of evaluation, thereby aiming to integrate the PES with control and planning systems, since the PES themselves are directed primarily towards planning, control and coordination of the production activities; and secondarily towards involvement of human resources.	Quantitative survey based approach based on principal component analysis with the aim of describing the dimensions and the actual state of PES.
Chan et al. (2001)	Investment appraisal, production system	The study presents theoretical review of technology investment justification methodologies and divides these approaches as strategic, economic, and analytic models. From the review it proposes a model to investments in technology, which is based around strategic planning, justification, training and installation, and routinization and implementation. It discusses issues related to each of these steps and offers ways to avoid and/or solve those problems. According to the study the performance variables that need to be considered include, performance, cost, quality, delivery, and innovativeness.	Qualitative review of research literature to provide ex ante performance measures and evaluation approaches for standalone systems, intermediate systems, and integrated systems.
Gupta and Whitehouse (2001)	Manufacturing strategy; manufacturing Technology	Presents a three dimensional (i.e. intensity of manufacturing technologies, organisational size, performance) PES, and provides the results of research carried out in manufacturing industries to examine the advanced manufacturing strategy based on organizational size. It concludes that large and small firms respond differently to technological changes.	Regression model base approach with data from 101 manufacturing organisations.
Lau et al. (2001)	Manufacturing management; outsourcing	Provides an OLAP based performance evaluation methodology to support enterprise decisions related to selecting appropriate business partners, by highlighting the importance of selecting the right partner for the right task.	Quantitative approach that utilises a scoring system for performance measures of quality, and delivery.

Continued next page

Authors	Scope	Description	Methodology Employed
Bitichi et al. (2002)	Manufacturing flexibility; strategic decision support	This research recognises the issue that in most evaluations performance data used is historic, and the process of data collection is time consuming. It presents an online system that collects performance data in real time from various manufacturing data sources such as ERP, customised spreadsheets and databases, and control systems to produce Shewhart charts to reveal performance of the area under investigation. The study concludes that appropriately designed and IT supported PES improves visibility, communications, teamwork, and decision making and aid proactive management.	Statistical analysis using an integrated performance measurement system (Bitichi and Carrie 1998), which identifies performance indicators and measures.
Kennerley and Neely (2002)	Process, people, systems, culture	This study investigates the issue of constituting organisation's performance measurement system, and argues that while formulating PES consideration is given to the current evaluation requirements of the organisation, rather than the emergent requirements which provides basis for a dynamic, change savvy measurement system. It provides empirical evidence of barriers and facilitators from 7 organisations that may impeded or aid the construction of a dynamic performance evlaution system.	Qualitative multiple case study approach involving semi structured interviews with 25 managers from a range of manufacturing organisations.
Ahmed (2002)	Resources utilisation; manufacturing management, learning process	This research examines theoretical frameworks in the area of performance measurement, and provides commentary on integrating performance measurement with business strategies. It proposes a computer based system that helps an organisation to choose an appropriate PES from a selection of 10 PES, depending upon the scope of evaluation sought.	Qualitative analysis of performance evaluation models, frameworks and systems, such as balanced scorecard and EFQM, to develop an IT enabled PES.
Das and Patel (2002)	Manufacturing systems; manufacturing flexibility; process planning	Discusses the why, what, and where of flexible manufacturing, and presents a performance evaluation tool for manufacturing managers to assess uncertainties in the manufacturing processes. The tool exploits the knowledge of employees to generate answers to the change issues; links operational changes to performance effects and then to manufacturing flexibility types; and prioritizes the changes so as to allow a step by step approach.	Structured questionnaire to identify and prioritise changes being experienced by the facility.
Yurdakul (2002)	Manufacturing system; competitive strategy; manufacturing quality, cost, dependability, flexibility.	This study examines the performance evaluation of manufacturing systems from a multi dimensional point of view. It argues that in a manufacturing PES performance criteria are interrelated and when the level of a single system criterion is changed, other criteria must be reconsidered simultaneously. It, thus, proposes an evaluation model that provides interrelated performance criteria for manufacturing systems. This PES takes into consideration manufacturing strategy and the interdependent performance criteria in its hierarchical structure.	Ex post evolution of the manufacturing system using a PES based on analytical hierarchy process and system with feedback approaches.

Continued next page

Authors	Scope	Description	Methodology Employed
Small and Yasin (2003)	Management information systems; manufacturing technology	This study examines the advanced manufacturing technology literature to develop a conceptual framework that demonstrates the influence of management information systems on the different facets of advanced manufacturing technologies adoption. The study bases its investigation on 12 performance measures, i.e. ability to change production lot sizes; variety of part-types or products manufactured; average number of tasks performed; operator output rates; revenue from manufacturing operations; deliver lead times; overhead costs; product quality; inventory turnover rates; production changeover times; time needed for major change in an existing product; and time to market for new products. It concludes that involvement of IT department in seeking technological solutions has positive impact on performance.	Survey among 125 corporate managers with the responsibility of evaluating technology solutions for manufacturing management.
Wongrassamee et al. (2003)	Business management; benchmarking; continuous mprovement.	The study provides a comparison of two improvement models i.e. balanced scorecard, and the EFQM excellence model. It concludes that both these approaches stem from similar concepts. It further concludes that it is difficult to find a perfect match between a company and a performance measurement framework, and the success of a measurement mechanism depends on its implementation.	Qualitative analysis from a critical perspective with regard to issues represented by questions relating to objectives, strategies, target setting, reward structures, and information feedback loops.
Kathuria and Porth (2003)	Manufacturing management; business strategy; business management.	This study provides an integrated view of strategy-managerial relationship characteristics and extends it the functional level of the organisation. It concludes that organisations with a higher level of strategy-managerial alignment at both the corporate level and business unit levels have a higher level of organisational performance. It also concludes that manufacturing units pursuing dissimilar strategies are driven by manufacturing managers with dissimilar attributes.	Quantitative approach based on survey of 196 managers within 98 organisations in the U.S.
Kuwaiti (2004)	Process reengineering; financial reporting; decision support	Examines the process of developing and managing performance evaluation systems. It concludes that the success of an evaluation system depends upon the reporting of performance data to the highest management level, and carrying out evaluation practices in collaboration with other business processes.	Survey based exploratory study among the members of institute of business process re-engineering.
Kumar and Harms (2004)	Process management; production improvement; continuous improvement; process planning	This study examines the challenges of corporate growth and profitability in manufacturing organisations. It provides a learning based approach through application of techniques such as process mapping and kaizen to improve basic business practices in a manufacturing organisation. The study demonstrates realisation of measurable results through use of process mapping tools, kaizen blitz activities, formalized and documented work instructions and work measurement tools.	Qualitative interpretive analysis of quantitative and qualitative user performance measures and controls to effectively manage operations.

Continued next page

Authors	Scope	Description	Methodology Employed
Tangen (2004)	Performance measures; activity based costing.	Presents a critique of contemporary approaches to performance measurement and attempts to identify whether they have in fact addressed the limitations of traditional ways of measuring performance. It concludes that contemporary frameworks indeed address conceptual issues; however they fall short of fulfilling the needs of actual industry practitioners.	Qualitative review of financial oriented, productivity/efficiency oriented, and multi dimensional performance evaluation systems.
Gosselin (2005)	Manufacturing strategy; production systems; materials management; production quality	Examines the extent to which manufacturing organisations employ performance mechanisms that are internally, as well as externally focused to include factors such as strategy and organizational structures. It concludes that manufacturing organisations continue to employ primarily financial performance measures in their measurement mechanisms, despite the recommendation from academics and industry experts. It further concludes that there are significant relationships between the types of measures and contextual factors like strategy, decentralization and environmental uncertainty.	Survey of 200 Canadian manufacturing organisations, data thus obtained analysed using a ranking approach.
Raymond (2005)	Advanced manufacturing technologies; information processing capability	Takes an information processing perspective for implementation of advanced manufacturing technologies. Using contingency theory perspective, the study determines the performance outcomes of the alignment between the critical success factors and their level of proficiency in the use of AMT. It concludes that critical success factors (such as quality, flexibility and information processing capability) and manufacturing technologies have a direct relationship with organisational performance in terms of operational efficiency, quality, cost reduction, productivity, and flexibility.	Survey of 118 Canadian manufacturers to determine the performance outcomes of the alignment.
Tapinos et al. (2005)	Performance measures; business strategy; organizational development	Examines impact of performance evaluation on strategic planning processes. It proposes performance evaluation as one of the four pillars of strategic planning and concludes that organisational size and rate of change in the business environment affects performance evaluation in strategic planning. It further concludes that large organisations and organisations operating in rapidly changing environments make greater use of performance evaluations.	Closed ended online survey based on Dyson's (2004) model to map the current practice of strategic planning and its influential factors.
Diaz et al. (2005)	Manufacturing technologies; operations management	Presents a case of implementation of advanced manufacturing technologies, by examining contemporary ex ante evaluation techniques. It concludes that use of financial measures in appraisal studies alone does not bring desirable advantages from implementation of advanced manufacturing technologies. It also concludes that to gain practical advantages from an evaluation system, variables that impact performance should be determined.	Survey and structured interviews at 20 asset installations, and the unit of analysis maintained through triangulation of data sources.

Continued next page

Authors	Scope	Description	Methodology Employed
Pegels and Watrous(2005)	Manufacturing systems; production management	Presents a study of an application of the theory of constraints to a manufacturing plant operations management environment. It proposes that to overcome operations issues and bottlenecks a multifaceted approach should be adopted.	Case study of the bottlenecks in the manufacturing process.
Bititci et al. (2006)	Organizational culture; management styles; organisational dynamics.	Investigates the relationship between organisational culture, management style, and evaluation systems. It concludes that this relationship is essential for effective management, and that the relationship between organisational culture and management style is interdependent, and collectively they impact on a measurement mechanism throughout its lifecycle. It further concludes that correct use of performance evaluation can lead to a cultural change.	Performance evaluation system implemented in five longitudinal case studies in action research programmes by identical implementation methods.
Parida and Kumar (2006)	Maintenance; performance measures, employee involvement; reliability; productivity.	Examines the issues relating to the challenges posed to development and implementation of an effective maintenance performance evaluation system. It concludes that for successful design and implementation of such a system it is essential to involve all staff associated with maintenance; the traditional overall equipment effectiveness methodology utilised by manufacturing organisations for maintenance management is inadequate; and in order to measure total maintenance effectiveness, it is essential to take into account both internal and external effectiveness.	An analytical approach in literature review to identify issues and challenges associated with maintenance performance measures.
Nachiappan and Anantharaman (2006)	Productive maintenance; line management; benchmarking.	Presents an approach to measure the overall line effectiveness in continuous line manufacturing systems. It utilises simulation approach to evaluate the effectiveness of continuous line manufacturing and identifies the bottlenecks and effects of specific contributing parameter for improvement. The result of this research makes it possible to represent the overall product line effectiveness as a benchmark for world class manufacturing to compare the performance of the various continuous product line manufacturing based industries.	Systematic methodology based on overall equipment effectiveness metrics to model the productivity of a line manufacturing system.
Gomes et al. (2006)	Production quality; customer satisfaction; process efficiency; quality management.	Investigates the issues relating to performance evaluation systems in manufacturing organisations. It concludes that, there is a consistent pattern of information on performance evaluations in manufacturing organisations regarding utilization, relevance and availability of performance measures; and that in a global manufacturing environment a multifaceted or multi dimensional evaluation technique is required.	Survey among 92 Portuguese manufacturing executives and data analysed using multiple regression analysis, cluster analysis and gap analysis.

Continued next page

Authors	Scope	Description	Methodology Employed
Garg and Deshmukh (2006)	Maintenance; maintenance optimization; information systems	Evaluates literature on maintenance management and highlights the gaps between researchers and practice. It concludes that there are a range of important issues relating to maintenance management ranging from optimization models to maintenance techniques, from scheduling of maintenance to information systems. It canvasses for a new approach to maintenance paradigm.	Qualitative content analysis.

Appendix D

Case Study Protocol

Interview Questions

1. What derives the strategic information systems plan in your organization? How often is it updated?

2. What are the drivers of IS implementation for asset management?

3. How do you regard IS adoption in your organization, as an enabler or an inhibitor of change? Do you have a clear policy with regard to IS adoption? For example, pioneer, adopter, follower, or laggard?

4. What has been the impact of IS on the productivity of employees as well as asset management processes?

5. What arrangements does the organization have for the review and audit of its IS supporting asset management to ensure risks are sufficiently mitigated and controls are in place to support asset management process?

6. Does your organization conduct risk assessments covering the organization's use of information technology, including internal systems and processes, outsourced services and the use of third-party communications and other services? If it does, how are the results acted on?

7. What procedures are in place to employee IS training?

8. How does your organization ensure that the stakeholders keep abreast of changes in business needs, technology, policies, and regulations?

9. What is your overall view of the technology being used for asset management?

10. What would you do differently if you were to implement IS for asset management again?

11. What types of computer based systems does your organisation employ at the moment?

Survey Questionnaire

Scale

1	-	**Non Existent**	:	Not present
2	-	**Initial/Ad hoc**	:	Elementary; replete with issues and problems
3	-	**Defined**	:	Controlled state; however still has issues
4	-	**Managed**	:	Managed state; however does not provide added value
5	-	**Optimized**	:	Economical, value added, and continuously improving

Description of Evaluation Criteria

Technical : Technical capability of IS infrastructure including hardware components (such as computers, communication networks, handheld devices, sensing devices, field devices and other tools) and software to perform required tasks.

Skills : Skills of employees in operating hardware and software

Information Availability : Does the organisation capture such information, and is it available to its stakeholders

Information Quality : What is the quality of information i.e. accuracy, timeliness, and completeness etc.

DESIGN PERSPECTIVE

How would you rate the performance of the following processes against the evaluation criteria and the scale defined above	Technical	Skills	Information Availability	Information Quality
Project management for asset construction or assembly				
Design of assets and facilities				
Asset and/or site layout design and schematic diagrams/drawings				
Asset bill of materials				
Naming and numbering of assets for configuration management				
Using existing data for operational, maintenance, and support tasks				
Use of historical information or cross project learnings in asset design process				
Identification of risks posed to asset operation (environmental and operational)				
Analysis of maintainability and reliability design requirements				
Failure modes, effects and criticality identification				
Asset lifecycle support forecasting in terms of spares acquisition, suppliers identification, inventory management, maintenance execution etc.				
Identification of items and materials that can be salvaged or recycled when assets or components become unserviceable				
Contingency planning for asset breakdown				
Evaluation of Design Alternatives				
Maintenance Task Analysis				
Reliability Analysis				
Level-of-Repair Analysis				
Maintenance Task Analysis				
Operator task analysis				
Design tradeoffs studies				
Prototype development				
Design review, test, and evaluation				
Acceptance testing				

OPERATIONS PERSPECTIVE

How would you rate the performance of the following processes against the evaluation criteria and the scale defined above	Technical	Skills	Information Availability	Information Quality
Developing asset operation workload schedule				
Responding to changes in production/service provision volume and asset operation				
Configuration management				
Balancing asset workload over the period of time to avoid troughs & peaks				
Operational environment testing				
Asset location and status Control				
Asset efficiency measurement				
Documentation support for asset operations (compliance with operating instructions and regulations, safety etc.)				
Asset operation compliance with environmental legislations				
Reduction or elimination of asset downtime and worker idleness by better matching throughputs of production/service provision				
Detection of production/service provision process defects				
Detection of reduced asset yield				
Asset operation risk assessment				
Asset condition monitoring and inspection				
Asset failure notifications (alarms etc.)				
Maintenance demand prediction and scheduling				
Maintenance work request generation				
Engineering control of hazardous material and substance				
Identification of standby assets in case of an asset failure				
Quality control and assurance during asset operation				
Asset performance and efficiency measurement				
Reduction in quality assurance non conformances				
Processes, tests, analysis and standards to be applied for the validation of asset performance				
Availability of quality policy and procedures throughout the organisation				
Communication of asset operation issues to other asset management functions				
Asset operation profiling				

MAINTENANCE PERSPECTIVE

How would you rate the performance of the following processes against the evaluation criteria and the scale defined above	Technical	Skills	Information Availability	Information Quality
Asset failure and wear patterns				
Maintenance needs analysis				
Failure modes effects and criticality analysis				
Routine maintenance work plan generation				
Emergency maintenance scheduling and follow up actions (such as disaster containment, maintenance resources allocation etc)				
Early signs of asset deterioration				
Maintenance work order system				
Establishing maintenance schedule				
Maintenance work request generation				
Asset shutdown scheduling				
Asset fault detection				
Asset fault root cause analysis				
Asset treatment options analysis (simulations etc.)				
Resource allocation for maintenance execution				
Spares logistics and inventory management				
Reduced variance in suppliers lead times and order handling				
Maintenance inventory and purchasing integration				
Maintenance workflow execution				
Documentation support for maintenance procedures				
Testing after servicing/repair treatment				
Identification of asset design weaknesses				
Asset remnant lifecycle calculation				
Reduction in maintenance expenses				
Maintenance/renewal trade offs				

OPERATIONAL EFFICIENCY PERSPECTIVE

How would you rate the performance of the following processes against the evaluation criteria and the scale defined above	Technical	Skills	Information Availability	Information Quality
Reduced process turnaround time				
Building integrated multidiscipline team for asset lifecycle management processes				
Handling customer complaints and asset operation safety non compliance incidents				
Reduction in operational skills required for asset operation and maintenance (i.e. Intelligent technologies doing it for humans)				
Integration of activities and processes with business partners				
Improved decision support				
Reduced unnecessary site visits				
Continuous process improvement				
Availability of on-job learning resources throughout the organisation				
Availability of written learning programs for multi-skilling and autonomous working				
Implementation of business improvement initiatives, such as Total Quality Management, Total Productive Maintenance				
Business intelligence gathering mechanism (both internal and external)				
Availability of historic asset performance and operational knowledge				
Documentation support for asset life cycle management process execution				
Communication of asset reliability assurance and need forecasting to process stakeholders				
Availability of information on vulnerabilities of asset operation to disasters and safety considerations				
Ability to audit process performance and assigning accountability				
Availability of information on industry best practices				
Informal communication networks within the organisation				

COMPETITIVENESS PERSPECTIVE

How would you rate the performance of the following processes against the evaluation criteria and the scale defined above	Technical	Skills	Information Availability	Information Quality
Understanding of the asset management issues faced by industry				
Availability to provide analysis on asset life cycle information for better planning and management in future				
Maintenance cost benefit analysis				
Asset lifecycle cost benefit analysis				
Asset obsoleteness and retirement recommendation				
Return on investment (ROI) analysis				
Assessment of existing approaches to operational efficiency				
Identification of underperforming areas to improve productivity				
Identification of market trends through analytical tools				
Facilitating coordinated asset management strategies				
Economic feasibility of asset management processes to the business				
Knowledge of the asset acquisition, expansion, and divestment strategies of competitors				
Asset lifecycle decision support				
Development and updating of asset management plan				
Increased integration of asset management processes				
Increased responsiveness to changes to asset management				
Innovations in asset design, operation, maintenance and support				
Enhanced integration with business partners				
Compliance with regulatory and environmental regulations				
Development of core competencies for asset management				
Third party services management				
Third party service level agreement management				
Customer feedback and satisfaction measurement				
Integration of inter organisational applications and processes				
Collaboration, assistance, and communication with stakeholders (i.e. business partners, emergency services, regulatory agencies)				
Matching customer needs with asset demand				
Establishing long term partnerships with key suppliers				
Sharing common coding and databases with business partners				
Sharing forecasting information with suppliers				
Sharing asset lifecycle data with contractors				

GPSR Compliance
The European Union's (EU) General Product Safety Regulation (GPSR) is a set
of rules that requires consumer products to be safe and our obligations to
ensure this.

If you have any concerns about our products, you can contact us on

ProductSafety@springernature.com

In case Publisher is established outside the EU, the EU authorized
representative is:

Springer Nature Customer Service Center GmbH
Europaplatz 3
69115 Heidelberg, Germany